T0331191

Materials and Strategies for Lab-on-a-Chip — Biological Analysis, Cell-Material Interfaces and Fluidic Assembly of Nanostructures

MATERIALS RESEARCH SOCIETY
SYMPOSIUM PROCEEDINGS VOLUME 1191

Materials and Strategies for Lab-on-a-Chip — Biological Analysis, Cell-Material Interfaces and Fluidic Assembly of Nanostructures

Symposium held April 14–17, 2009, San Francisco, California, U.S.A.

EDITORS:

Shashi K. Murthy
Northeastern University
Boston, Massachusetts, U.S.A.

Saif A. Khan
National University of Singapore
Singapore

Victor M. Ugaz
Texas A&M University
College Station, Texas, U.S.A.

Henry C. Zeringue
University of Pittsburgh
Pittsburgh, Pennsylvania, U.S.A.

Materials Research Society
Warrendale, Pennsylvania

CAMBRIDGE
UNIVERSITY PRESS

University Printing House, Cambridge CB2 8BS, United Kingdom

One Liberty Plaza, 20th Floor, New York, NY 10006, USA

477 Williamstown Road, Port Melbourne, VIC 3207, Australia

314-321, 3rd Floor, Plot 3, Splendor Forum, Jasola District Centre, New Delhi - 110025, India

79 Anson Road, #06-04/06, Singapore 079906

Cambridge University Press is part of the University of Cambridge.

It furthers the University's mission by disseminating knowledge in the pursuit of education, learning and research at the highest international levels of excellence.

www.cambridge.org
Information on this title: www.cambridge.org/9781605111643

Materials Research Society
506 Keystone Drive, Warrendale, PA 15086
http://www.mrs.org

First published 2009
First paperback edition 2012

Single article reprints from this publication are available through University Microfilms Inc., 300 North Zeeb Road, Ann Arbor, MI 48106

CODEN: MRSPDH

A catalogue record for this publication is available from the British Library

ISBN 978-1-605-11164-3 Hardback
ISBN 978-1-107-40815-9 Paperback

FRONTIERS IN LAB-ON-A-CHIP RESEARCH

POSTER SESSION: MATERIALS FOR LAB ON A CHIP

MATERIALS SYNTHESIS ON CHIP

*Invited Paper

CELL MANIPULATION AND BIOMIMETICS ON CHIP

ADVANCES IN DEVICE MATERIALS

ADVANCES IN INTEGRATING DEVICE COMPONENTS

POROUS MATERIALS IN LABS ON A CHIP

SENSING AND DETECTION ON CHIP — MOLECULAR LEVEL

SENSING AND DETECTION ON CHIP — CELLS AND PARTICLES

PREFACE

This volume is a collection of papers presented at Symposium OO, "Materials and Strategies for Lab-on-a-Chip — Biological Analysis, Cell-Material Interfaces and Fluidic Assembly of Nanostructures," held April 14–17 at the 2009 MRS Spring Meeting in San Francisco, California.

The development of miniaturized systems for chemical and biochemical analysis has grown and matured to the point where lab-on-a-chip devices are now important enabling tools in a diverse array of application areas. But as the size of these systems continues to shrink, details of the micro- and nano-scale phenomena associated with their construction and operation must be considered. This symposium addressed these challenges by assembling a unique and vibrant collection of speakers who presented multidisciplinary research at the interface between materials science, biology, chemistry, and engineering.

We are grateful to all of the contributors and participants who made this four-day symposium highly successful. We gratefully acknowledge financial support from Microfluidic ChipShop GmbH and Northeastern University.

Shashi K. Murthy
Saif A. Khan
Victor M. Ugaz
Henry C. Zeringue

August 2009

MATERIALS RESEARCH SOCIETY SYMPOSIUM PROCEEDINGS

Volume 1153 — Amorphous and Polycrystalline Thin-Film Silicon Science and Technology — 2009, A. Flewitt, Q. Wang, J. Hou, S. Uchikoga, A. Nathan, 2009, ISBN 978-1-60511-126-1
Volume 1154 — Concepts in Molecular and Organic Electronics, N. Koch, E. Zojer, S.-W. Hla, X. Zhu, 2009, ISBN 978-1-60511-127-8
Volume 1155 — CMOS Gate-Stack Scaling — Materials, Interfaces and Reliability Implications, J. Butterbaugh, A. Demkov, R. Harris, W. Rachmady, B. Taylor, 2009, ISBN 978-1-60511-128-5
Volume 1156— Materials, Processes and Reliability for Advanced Interconnects for Micro- and Nanoelectronics — 2009, M. Gall, A. Grill, F. Iacopi, J. Koike, T. Usui, 2009, ISBN 978-1-60511-129-2
Volume 1157 — Science and Technology of Chemical Mechanical Planarization (CMP), A. Kumar, C.F. Higgs III, C.S. Korach, S. Balakumar, 2009, ISBN 978-1-60511-130-8
Volume 1158E —Packaging, Chip-Package Interactions and Solder Materials Challenges, P.A. Kohl, P.S. Ho, P. Thompson, R. Aschenbrenner, 2009, ISBN 978-1-60511-131-5
Volume 1159E —High-Throughput Synthesis and Measurement Methods for Rapid Optimization and Discovery of Advanced Materials, M.L. Green, I. Takeuchi, T. Chiang, J. Paul, 2009, ISBN 978-1-60511-132-2
Volume 1160 — Materials and Physics for Nonvolatile Memories, Y. Fujisaki, R. Waser, T. Li, C. Bonafos, 2009, ISBN 978-1-60511-133-9
Volume 1161E —Engineered Multiferroics — Magnetoelectric Interactions, Sensors and Devices, G. Srinivasan, M.I. Bichurin, S. Priya, N.X. Sun, 2009, ISBN 978-1-60511-134-6
Volume 1162E —High-Temperature Photonic Structures, V. Shklover, S.-Y. Lin, R. Biswas, E. Johnson, 2009, ISBN 978-1-60511-135-3
Volume 1163E —Materials Research for Terahertz Technology Development, C.E. Stutz, D. Ritchie, P. Schunemann, J. Deibel, 2009, ISBN 978-1-60511-136-0
Volume 1164 — Nuclear Radiation Detection Materials — 2009, D.L. Perry, A. Burger, L. Franks, K. Yasuda, M. Fiederle, 2009, ISBN 978-1-60511-137-7
Volume 1165 — Thin-Film Compound Semiconductor Photovoltaics — 2009, A. Yamada, C. Heske, M. Contreras, M. Igalson, S.J.C. Irvine, 2009, ISBN 978-1-60511-138-4
Volume 1166 — Materials and Devices for Thermal-to-Electric Energy Conversion, J. Yang, G.S. Nolas, K. Koumoto, Y. Grin, 2009, ISBN 978-1-60511-139-1
Volume 1167 — Compound Semiconductors for Energy Applications and Environmental Sustainability, F. Shahedipour-Sandvik, E.F. Schubert, L.D. Bell, V. Tilak, A.W. Bett, 2009, ISBN 978-1-60511-140-7
Volume 1168E —Three-Dimensional Architectures for Energy Generation and Storage, B. Dunn, G. Li, J.W. Long, E. Yablonovitch, 2009, ISBN 978-1-60511-141-4
Volume 1169E —Materials Science of Water Purification, Y. Cohen, 2009, ISBN 978-1-60511-142-1
Volume 1170E —Materials for Renewable Energy at the Society and Technology Nexus, R.T. Collins, 2009, ISBN 978-1-60511-143-8
Volume 1171E —Materials in Photocatalysis and Photoelectrochemistry for Environmental Applications and H_2 Generation, A. Braun, P.A. Alivisatos, E. Figgemeier, J.A. Turner, J. Ye, E.A. Chandler, 2009, ISBN 978-1-60511-144-5
Volume 1172E —Nanoscale Heat Transport — From Fundamentals to Devices, R. Venkatasubramanian, 2009, ISBN 978-1-60511-145-2
Volume 1173E —Electofluidic Materials and Applications — Micro/Biofluidics, Electowetting and Electrospinning, A. Steckl, Y. Nemirovsky, A. Singh, W.-C. Tian, 2009, ISBN 978-1-60511-146-9
Volume 1174 — Functional Metal-Oxide Nanostructures, J. Wu, W. Han, A. Janotti, H.-C. Kim, 2009, ISBN 978-1-60511-147-6

MATERIALS RESEARCH SOCIETY SYMPOSIUM PROCEEDINGS

Volume 1175E —Novel Functional Properties at Oxide-Oxide Interfaces, G. Rijnders, R. Pentcheva, J. Chakhalian, I. Bozovic, 2009, ISBN 978-1-60511-148-3
Volume 1176E —Nanocrystalline Materials as Precursors for Complex Multifunctional Structures through Chemical Transformations and Self Assembly, Y. Yin, Y. Sun, D. Talapin, H. Yang, 2009, ISBN 978-1-60511-149-0
Volume 1177E —Computational Nanoscience — How to Exploit Synergy between Predictive Simulations and Experiment, G. Galli, D. Johnson, M. Hybertsen, S. Shankar, 2009, ISBN 978-1-60511-150-6
Volume 1178E —Semiconductor Nanowires — Growth, Size-Dependent Properties and Applications, A. Javey, 2009, ISBN 978-1-60511-151-3
Volume 1179E —Material Systems and Processes for Three-Dimensional Micro- and Nanoscale Fabrication and Lithography, S.M. Kuebler, V.T. Milam, 2009, ISBN 978-1-60511-152-0
Volume 1180E —Nanoscale Functionalization and New Discoveries in Modern Superconductivity, R. Feenstra, D.C. Larbalestier, B. Maiorov, M. Putti, Y.-Y. Xie, 2009, ISBN 978-1-60511-153-7
Volume 1181 — Ion Beams and Nano-Engineering, D. Ila, P.K. Chu, N. Kishimoto, J.K.N. Lindner, J. Baglin, 2009, ISBN 978-1-60511-154-4
Volume 1182 — Materials for Nanophotonics — Plasmonics, Metamaterials and Light Localization, M. Brongersma, L. Dal Negro, J.M. Fukumoto, L. Novotny, 2009, ISBN 978-1-60511-155-1
Volume 1183 — Novel Materials and Devices for Spintronics, O.G. Heinonen, S. Sanvito, V.A. Dediu, N. Rizzo, 2009, ISBN 978-1-60511-156-8
Volume 1184 — Electron Crystallography for Materials Research and Quantitative Characterization of Nanostructured Materials, P. Moeck, S. Hovmöller, S. Nicolopoulos, S. Rouvimov, V. Petkov, M. Gateshki, P. Fraundorf, 2009, ISBN 978-1-60511-157-5
Volume 1185 — Probing Mechanics at Nanoscale Dimensions, N. Tamura, A. Minor, C. Murray, L. Friedman, 2009, ISBN 978-1-60511-158-2
Volume 1186E —Nanoscale Electromechanics and Piezoresponse Force Microcopy of Inorganic, Macromolecular and Biological Systems, S.V. Kalinin, A.N. Morozovska, N. Valanoor, W. Brownell, 2009, ISBN 978-1-60511-159-9
Volume 1187 — Structure-Property Relationships in Biomineralized and Biomimetic Composites, D. Kisailus, L. Estroff, W. Landis, P. Zavattieri, H.S. Gupta, 2009, ISBN 978-1-60511-160-5
Volume 1188 — Architectured Multifunctional Materials, Y. Brechet, J.D. Embury, P.R. Onck, 2009, ISBN 978-1-60511-161-2
Volume 1189E —Synthesis of Bioinspired Hierarchical Soft and Hybrid Materials, S. Yang, F. Meldrum, N. Kotov, C. Li, 2009, ISBN 978-1-60511-162-9
Volume 1190 — Active Polymers, K. Gall, T. Ikeda, P. Shastri, A. Lendlein, 2009, ISBN 978-1-60511-163-6
Volume 1191 — Materials and Strategies for Lab-on-a-Chip — Biological Analysis, Cell-Material Interfaces and Fluidic Assembly of Nanostructures, S. Murthy, H. Zeringue, S. Khan, V. Ugaz, 2009, ISBN 978-1-60511-164-3
Volume 1192E —Materials and Devices for Flexible and Stretchable Electronics, S. Bauer, S.P. Lacour, T. Li, T. Someya, 2009, ISBN 978-1-60511-165-0
Volume 1193 — Scientific Basis for Nuclear Waste Management XXXIII, B.E. Burakov, A.S. Aloy, 2009, ISBN 978-1-60511-166-7

Prior Materials Research Society Symposium Proceedings available by contacting Materials Research Society

Frontiers in Lab-on-a-Chip Research

Mater. Res. Soc. Symp. Proc. Vol. 1191 © 2009 Materials Research Society 1191-OO02-01

Microfluidic Manifolds With High Dynamic Range in Structural Dimensions Replicated in Thermoplastic Materials

Holger Becker[1], Erik Beckert[2], Claudia Gärtner[1]

[1] microfluidic ChipShop GmbH, Carl-Zeiss-Promenade 10, D-07745 Jena, Germany
[2] Fraunhofer Institute for Applied Optics and Precision Engineering, Albert-Einstein-Str. 7, D-07745 Jena, Germany

ABSTRACT

In this paper, we present the manufacturing process of a polymer microfluidic device which is currently being used to investigate wetting properties of nanostructured microchannels replicated in hydrophobic thermoplastic materials like cyclo-olefin co-polymer (COC), polypropylene (PP) or polymethylmetacrylate (PMMA). These devices feature large structural dynamics (feature sizes between 200 μm and 200 nm). The mold insert necessary was fabricated using a combination of precision machining with single-point diamond turning (SPDT).

INTRODUCTION

Microfluidic components nowadays play a major role in ensuring the performance of many systems in analytical chemistry and the life sciences [1]. In many real-world applications, a set of requirements have to be met which differs significantly from properties reported in the literature by many academic groups. Amongst those requirements are:

1. A high dynamic range in structural dimensions. As an example, in a diagnostic cartridge for full blood analysis with external dimensions typically the size of a microscopy slide (75.5 × 25.5 mm), initial sample and reagent volumes are typically of the order of several μl (i.e. several mm^3), requiring chambers with dimensions in the mm-range. The various liquids are then transported and manipulated in microchannels which are typically in the size range of tens to hundreds of micrometers. Specific structures (e.g. obstacles to support mixing, cell or bead capture, passive valves) and the general tolerances are often one order of magnitude smaller (e.g. 1-10 μm), while structures influencing the surface properties (e.g. wetting) are in the submicron range. A single microfluidic device will incorporate structures in all these size ranges in all three spatial dimensions which generally is not possible for manufacturing methods based on lithographic techniques.
2. To enable commercialization of such a microfluidic device, the manufacturing technology used has to be scalable in the sense that it allows the fabrication of devices from low to mass production volumes at reasonable cost (typically for an analytical or diagnostic test in the low single-digit dollar range) without a switch in the basic technology.
3. Material selection has to be compatible with both the manufacturing technologies and the application needs in addition to meeting the cost restraints indicated above.

In order to fulfil these requirements, we have investigated the replication of microscopy slide sized micro- and nanostructured metallic mold inserts into thermoplastic materials like

PMMA, PP and COC using hot embossing and injection molding as manufacturing technologies. These replication technologies are established for the fabrication of macroscopic structures and have been increasingly used to generate microstructures as well [2]. However the combination of comparatively workpieces with a large dynamic range of structures has neither been subject to extensive academic research nor industrial manufacturing yet.

EXPERIMENT

Mold insert fabrication

A crucial step in the technology chain for the manufacturing of polymer components is the fabrication of a suitable mold insert (or mold master) structure which contains the inverse structure of the final polymer part. While electroplating of silicon or resist structures has been used frequently to generate a metallic (usually nickel or a nickel alloy) mold insert with features in the micrometer-range, this technology is usually not well suited for structures with a height larger than several hundred micrometers. Silicon [3] or resists [4] structures have been used directly as mold insert materials, however in addition to the above mentioned geometrical restraints their lifetime is significantly smaller than that of a metallic mold insert.
From the perspective of lifetime, metallic mold inserts are very attractive. The advances of precision- and ultraprecision mechanical machining in recent years have made it possible to achieve the required high dynamic range in geometrical features combined with the long mold insert lifetime required for economic viability of the method.
While ultraprecision machining using diamond tools had its origins in optical component manufacturing [5], e.g. large mirrors or grating structures for antireflex functionality [6], it can also be used for the manufacturing of mold inserts for polymer replication applications like microfluidic.
The mold insert for the microfluidic channel structure with an overlayed saw-tooth at the channel bottom (which is equivalent to a ridge top in the mold insert) was manufactured using a precision milling machine (Kern Evo Ultra-precision 5-axis machine center, Kern, Eschenlohe, Germany) to generate the microstructured channel (100 and 200 µm wide microchannels, 50 µm deep, overall layout see Fig. 1) and ultra precision single point diamond turning (SPDT) to generate the nanostructured channel bottom. On a turning machine (Nanoform 350, Precitech Inc., Keene, USA), two work pieces were arranged on an outmost diameter of the turning spindle, thus approximating a linear cut through the channel with a width of 100 µm and 200 µm respectively by the large turning radius. The working tool was a mono crystalline diamond cutting tool (see Fig. 2) with a tip shape adapted to the targeted saw tooth structure (2 µm pitch and 0.3 µm depth). Manufacturable materials for diamond tools are most non-ferrous metals as well as most polymers and some amorphous crystallic materials such as silicon. In case of the mould master for the channel structure a brass alloy of $CuZn_{39}Pb_2F_{38}$ was chosen due to its

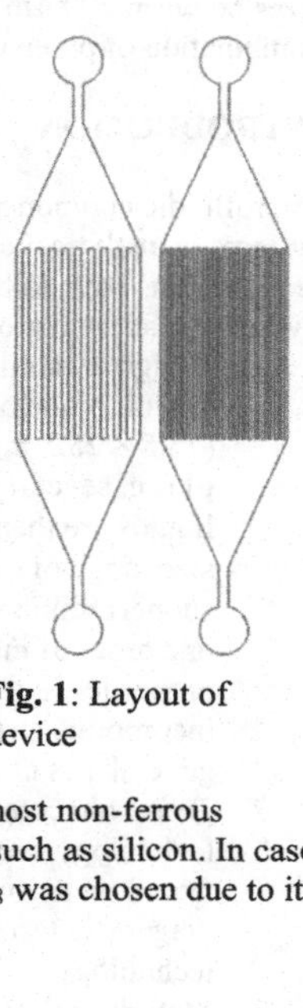

Fig. 1: Layout of device

good machineability. For production tools, harder materials such as electroplated NiP on steel bases are available.

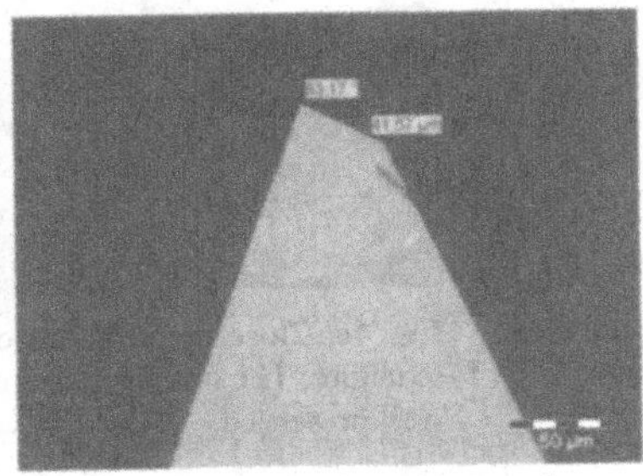

Fig. 2: Diamond cutting tool

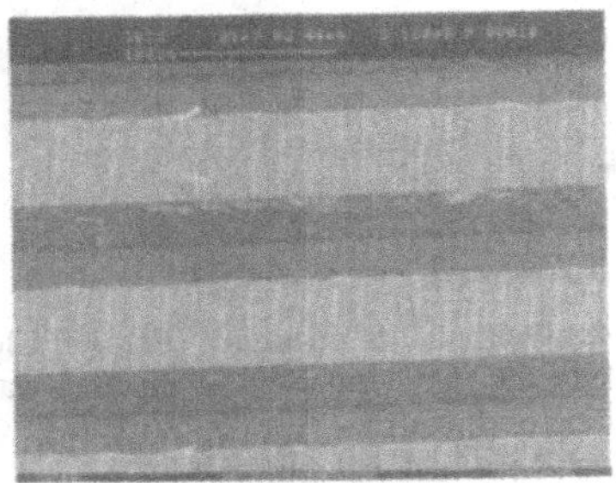

Fig. 3: SEM of mold insert

Additionally, it becomes more and more evident that cubic boron nitride (CBN) will be available in future within the same range of accuracies and geometries as diamond tools, thus eliminating the tool wear effect due to carbon affinity of diamond and allowing the ultra precision machining of steel moulds.
Fig. 3 shows a SEM-view of the manufactured brass mold insert. Visible are the milling structures from conventional machining as well as the saw tooth structure on the channel bottom. The machining degrees of freedom allow for different angles of the grating with respect to the flow direction as well as the structuring of more complex geometries such as steps with different heights down to the sub micron scale.

Polymer replication

The above described mold insert was used for the replication in a variety of thermoplastic polymers by injection molding and hot embossing.
For injection molding (Figs. 4,5), as can be seen in the overview images (Fig. 4a and 5a), the replication of the microstructure is not perfect yet (rounding of the edges). The nanostructure is well defined in both materials, however with greater replication accuracy in the case of COC than in PP. This can be associated with the shorter polymeric chain length and the better mold flow in the case of COC as well as the higher mechanical stability after cooling down which leads to a reduced structural deformation upon demolding.

Fig. 4a: Overview of injection molded COC. Note the rounding of the micro-structured channel wall due to incomplete mold filling.

Fig. 4b: Close-up of the area channel-wall and nano-structure.

Fig. 4c: Close-up of the nano-structure. The dimensions are well preserved.

Experiments have also been carried out using the hot embossing process (Jenoptik Mikrotechnik HEX 02, Jena, Germany) which is known from the literature to have a very high replication accuracy [7,8]. This property could be confirmed both in COC and PMMA. Figures 6 and 7 show the replication results in these two materials, COC and PMMA. The master tool could be very well replicated, the microstructure (as indicated by the vertical sidewalls) as well as the nanostructure.

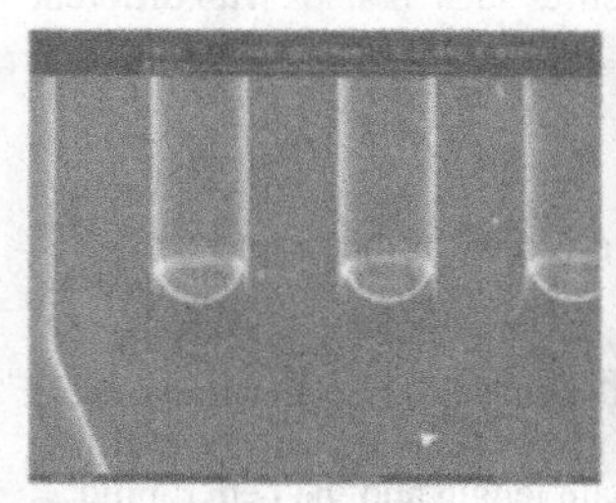

Fig. 5a: Overview of injection molded PP. Note the rounding of the micro-structured channel wall due to incomplete mold filling and some deformation at the ends due to demolding.

Fig. 5b: Close-up of a single channel with nanostructures.

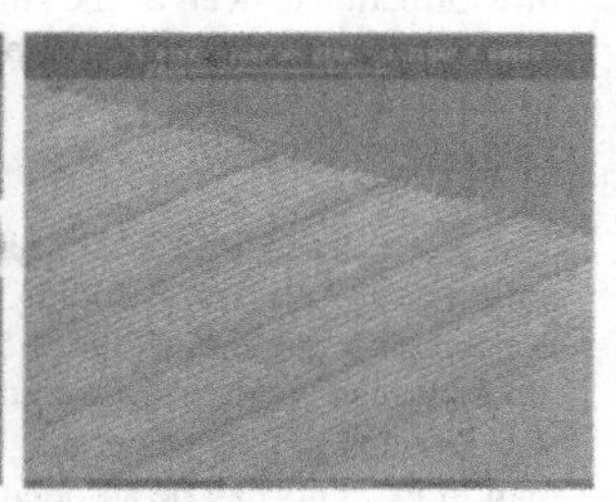

Fig. 5c: Close-up of the area channel-wall and nano-structure. The structures are not as well replicated as in the case of COC.

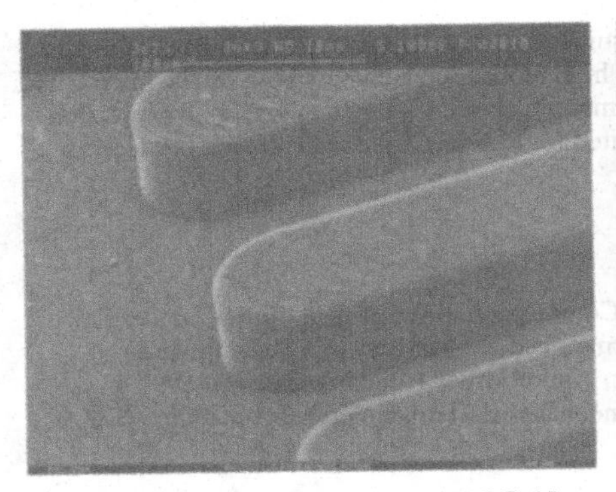

Fig. 6a: Hot embossed structure in COC. Note the high replication accuracy both of the micro- as well as the nanostructure.

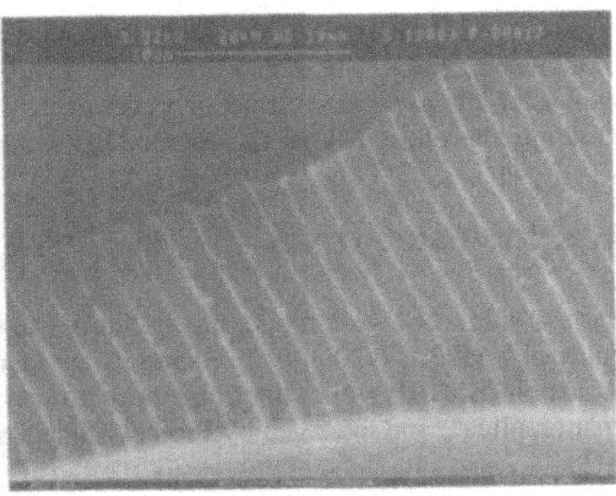

Fig. 6b: Detail of the nanostructure.

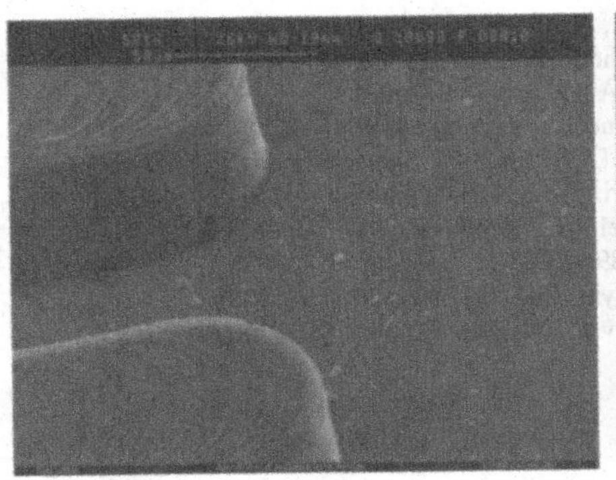

Fig. 7a: Hot embossed structure in PMMA. Note the high replication accuracy both of the micro- as well as the nanostructure.

Fig. 7b: Detail of the nanostructure. The resolution of the SEM could not be as high as in the case of COC as thermal damage in the PMMA cold be observed due to the electron beam.

CONCLUSIONS

We have been able to replicate a microfluidic channel device with a sawtooth-profiled nanostructure on the channel floor to allow wetting of the intrinsically hydrophobic polymer surface. The combination of precision milling with SPDT proves to be a suitable method for the generation of a mold insert for the subsequent replication technologies. Injection molding as the process which offers the biggest economic potential is well capable of replicating the nanostructure, albeit at the (possible) cost of a not optimally developed microstructure. It can be argued however that if this deviation is known, suitable corrective measures can be taken in the mold insert device. Hot embossing shows a very high replication accuracy both on the micron as well as on the nanometer scale, however cycle times are about 4-5 times as long as for injection molding, therefore making this method better suited for lower volume or prototype production.

ACKNOWLEDGMENTS

Parts of this work were funded by the European Commission in the project "Influs" under the contract NMP3-CT-2006-031980 and by the German Federal Ministry of Education and Research (BMBF) within the framework program "InnoProfile" (fund number 03IP609, "nanoreplica") and managed by Project Management Juelich, Forschungszentrum Jülich GmbH (PTJ).

REFERENCES

.

1. Nature Insight: Lab on a chip, **442**, 7101, 367-418 (2006).
2. H. Becker, C. Gärtner: Anal. Bioanal. Chem. **390**, 89-111, (2008).
3. H. Becker, U. Heim: Sensors Actuators A **83**, 130-135 (2000).
4. J. Narasimhan, I. Papautsky, I, J. Micromech Microeng **14**, 96–103, (2004).
5. A. Gebhardt, R. Steinkopf, in Proc. EUSPEN, 4th International Conference of the European Society for Precision Engineering and Nanotechnology, 197-198 (2004).
6. C. Brückner, B. Pradarutti, O. Stenzel, R. Steinkopf, S. Riehemann, G. Notni, A. Tünnermann, Optics Express **15**, 779–789, (2007).
7. Y.J. Juang, L.J. Lee, K.W. Koelling, Polym. Eng. Sci. **42**, 539–550 (2002).
8. Y.J. Juang, L.J. Lee, K.W. Koelling, Polym. Eng. Sci. **42**, 551–566 (2002).

Mater. Res. Soc. Symp. Proc. Vol. 1191 © 2009 Materials Research Society 1191-OO02-02

Tetherless, 3D, Micro-Nanoscale Tools and Devices for Lab on a Chip Applications

David H. Gracias [1,2]
[1]Department of Chemical and Biomolecular Engineering and [1]Department of Chemistry, Johns Hopkins University, 3400 N. Charles Street, Baltimore, MD 21218, USA

ABSTRACT

On the macroscale, a laboratory scientist uses a large number of tools such as flasks, crucibles, spatulas, filter funnels, test tube holders and grippers. If laboratory procedures are to be miniaturized within chips, micro and nanoscale analogs of these macroscopic tools could provide enhanced functionality. In this paper, I describe research efforts in our group aimed at engineering three dimensional, tetherless and lithographically patterned miniaturized structures for lab on a chip applications.

INTRODUCTION

There are numerous tetherless structures such as liquid droplets, magnetic and polymeric beads and molecular containers that have been widely utilized in lab on a chip systems to enable chemistry and chemical delivery with spatio-temporal control [1-2]. Mobile 3D encapsulation devices can be used to deliver chemicals, on-demand, to specific locations within the chip. Miniaturized grasping tools can be used to deliver, capture, and retrieve objects to-and-from hard to reach places within fluidic channels. As laboratory scale procedures are miniaturized, in addition to planar and quasi planar fluidic devices, there is a need to develop miniaturized analogs of macroscopic laboratory tools and devices. Over the last five years, our research group has utilized lithographic fabrication and self-assembly to fabricate these tools and related devices for lab on a chip applications.

RESULTS and DISCUSSION

There are several unique features of the devices we have constructed.

Lithographic Patterning

Our tools and devices have been patterned using photolithography and electron-beam lithography. Conventional lithographic patterning can enable highly precise patterning and monodisperse structuring of devices in a parallel and cost effective manner. Additionally, lithographic processes are compatible with the integration of electronic, optical and communication devices. Hence, by utilizing lithographic patterning, it may be possible to integrate these components onto mobile lab on a chip devices.

Self-assembly to enable three dimensionality

A severe limitation of many lithographic processes is that they are inherently two dimensional. Hence, we utilize self-assembly to structure lithographic components and integrate devices in 3D. The need for three dimensional patterning in miniaturized lab on a chip devices is evident from Fig. 1. Typically, while it is relatively straightforward to pattern devices along certain axes, it is far more challenging to pattern them along orthogonal axes. In the case of an encapsulant or microwell; limited patterning allows the encapsulated objects to interact with their surroundings only along one axis. In contrast, 3D patterning enables them to interact in all three dimensions. This interaction is important when the encapsulated objects are cells or living organisms, whose very existence depends on an exchange of chemicals with the surrounding media.

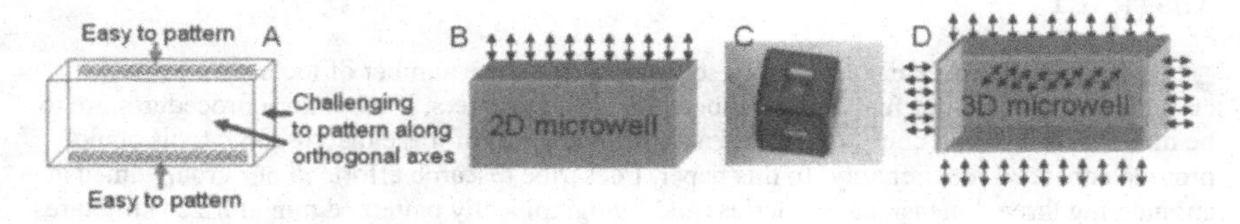

Figure 1. A. When objects are patterned using lithography, while it is relatively straight forward to pattern them along the direction of the substrate, it is challenging to pattern them along the orthogonal axes. **B.** A conventional 2D microwell allowing perfusion along one axis **C.** Optical image of a container fabricated in our laboratory, that is lithographically patterned in all three dimensions [3]; the container encapsulates a *Triops* embryo (image obtained by Jillian Epstein) **D.** Patterning in all three dimensions enables perfusion along all three axes.

<u>Two kinds of hinges</u>

Our structures consist of lithographically patterned panels connected by a series of hinges. The hinges tether the panels and keep them in close proximity, so that they can assemble correctly. The hinges also cause the panels to rotate which is critical in enabling the structure to fold up from a 2D geometry to a 3D one. We have utilized hinges that operate on two different mechanisms; one based on surface tension [4], the other based on stressed thin film bilayers [5]. In addition to hinges that cause the panels to rotate, there are "locking" hinges that fuse together to seal the structure at the edges. Maintaining robust leak-proof seals at the corners of these structures is a continuing area of research in our group, as is the development of hinges that can be triggered on-demand.

<u>Lithographically patterned containers (Fig. 2)</u>

We have developed containers that are three dimensionally patterned with side a length ranging from 100 nm to 2 mm. Side wall patterning has been demonstrated down to 15 nm. Containers with specific side wall patterning including nanoporosity in all three dimensions have been achieved [6]. Most containers have been constructed of metals such as gold and nickel; however, containers can be constructed with any rigid photopatternable material.

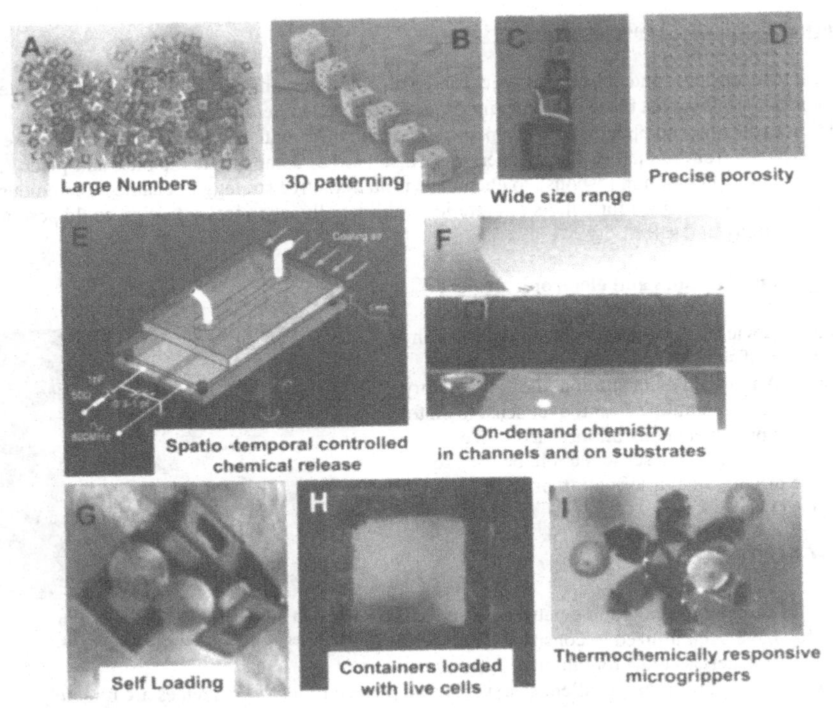

Figure 2. A-H: 3D mobile containers. A. Containers can be fabricated in large numbers **B.** They can have nanometer scale arbitrary patterning on one or all faces. **C.** Containers have been fabricated with a wide range of sizes **D.** Precise monodisperse side wall micro-nano porosity can be patterned. **E.** Containers can be moved by magnets and remotely heated to release chemicals by radio frequency waves. **F.** Hence, they can be used to carry out on-demand chemistry in channels or on substrates. **G.** Self-loading containers. **H.** Containers can encapsulate live cells and are not toxic. **I.** Thermochemically responsive microgrippers. Images taken by Timothy Leong, Christina Randall, Hongke Ye, and Jillian Epstein.

The containers with thin film hinges are self-loading and they can be filled with both living and non-living objects during assembly. Cells can be cultured in the containers after encapsulation [6-7]. Since the containers are magnetic and metallic they can be moved with magnets from distances as far away as several centimeters. They can also be remotely heated using radio frequency waves [8]. These features allow the containers to deliver chemicals with spatio-temporal control.

Thermochemically responsive microgrippers

The capture, retrieval and delivery of objects are valuable functions that have been demonstrated using microgrippers that close and open in response to both temperature and chemicals. In addition to closing and opening, the grippers can be structured with sharp claws that can be used to excise cells from tissue [9]. The grippers were also used to demonstrate the pick-and-place function [10]. Actuation in response to chemicals is an attractive strategy to enable autonomous function without the need for tethers or batteries. Additionally, chemical actuation could permit high specificity and selectivity.

Integration of sensors and electronic devices

Since the devices are constructed using lithography, integration of sensors and electronic modules is possible. An initial demonstration in our laboratory involved the integration of cantilever sensors on the containers patterned along all three axes. We demonstrated that such sensors could be used to measure orientational information of approaching analytes [11]

Figure 3 SEM image of containers with cantilever sensors patterned on all faces.

CONCLUSIONS

In summary, lithographic patterning and self-assembly have been utilized to construct a number of mobile, 3D tetherless tools for lab on a chip applications. Several future challenges exist. At the present time, these devices are remotely moved by a human operator; however, we are trying to enable autonomous motion based on flow, Brownian motion, solvent driven motion, chemotaxis and chemical reactions. There is also the need to fabricate such devices with a wide range of materials; as an example, it is difficult to image the contents of 3D containers using optical microscopy, and there is a need to construct transparent versions of such devices to view encapsulated cells in 3D. There is also a need to enable these devices to close and open in a multi-use reversible manner. Nevertheless, these challenges do not seem insurmountable.

ACKNOWLEDGMENTS

The research described in this paper was performed by a number of students including Timothy Leong, Christina Randall, Jeong-Hyun Cho, Hongke Ye, Zhiyong Gu, Bryan Benson, Emma Call, Anum Azam, Jillian Epstein and Aasiyeh Zarafshar as described in the references cited below.

REFERENCES

1. S.-Y Teh, R. Lin, L.-H. Hung, Lung-Hsin and A. P. Lee, Lab Chip, 8, 198 (2008).
2. A. M. Gijs, Microfluidics and Nanofluidics 1, 22 (2004).

3. T. G. Leong, C. L. Randall, B. R. Benson, A.M. Zarafshar and D. H. Gracias, Lab Chip 8,1621 (2008)
4. T. Leong, P. Lester, T. Koh, E. Call and D. H. Gracias, Langmuir 23, 8747 (2007)
5. T. G. Leong, B. R. Benson, E. K. Call, and D. H. Gracias, Small 4,1605 (2008)
6. C. L. Randall, A. Gillespie, S. Singh, T. G. Leong and D. H. Gracias, Anal. Bioanal. Chem. 393,1217–1224 (2009)
7. B. Gimi, T. Leong, Z. Gu, M.Yang, D. Artemov, Z. M. Bhujwalla and D. H. Gracias, Biomed. Microdevices 7, 341 (2005)
8. H. Ye, C. Randall, T. Leong, D. Slanac, E. Call and D. H. Gracias, Angew. Chem., Int. Ed. 46, 4991 (2007).
9. T. G. Leong, C. L. Randall, B. R. Benson, N. Bassik, G. M. Stern and D. H. Gracias, Proc. Natl. Acad. Sci. U. S. A. 106, 703 (2009)
10. J. S. Randhawa, T. G. Leong, N. Bassik, B. R. Benson, M. T. Jochmans and D.H. Gracias, J. Am. Chem. Soc.130, 7238 (2008).
11. J. H. Cho, S. Hu and D. H. Gracias, App. Phys. Lett. 93, 043505/1 (2008).

Poster Session:
Materials for Lab on a Chip

Mater. Res. Soc. Symp. Proc. Vol. 1191 © 2009 Materials Research Society 1191-OO03-01

Hot Embossing of Microchannels in Cyclic Olefin Copolymer

Patrick W. Leech[1]
[1]CSIRO Materials Science and Engineering, Clayton, 3168, Victoria, Australia

ABSTRACT

The hot embossing properties of Cyclic Olefin Copolymer (COC) have been examined as a function of comonomer content. Six standard grades of COC with varying norbornene content (61-82 wt%) were used in these experiments in order to provide a range of glass transition temperatures, T_g. All grades of COC exhibited sharp increases in embossed depth over a critical range of temperature. The transition temperature in embossed depth increased linearly with norbornene content for both 35 and 70 μm deep structures. At temperatures above this transition, the dimensions of the embossed patterns were essentially independent of COC grade, the applied pressure and duration of loading. Channels formed above the transition in a regime of viscous liquid flow were extremely smooth in morphology for all grades. The average surface roughness, R_a, measured at the base of the channels decreased sharply at the transition temperature, with a levelling off at higher temperatures. Grades of COC with higher norbornene content exhibited extensive micro-cracking during embossing at temperatures close to the transition temperature.

INTRODUCTION

Micro/nanofluidic systems based on polymers have become increasingly important in the realisation of portable devices for biomedical analysis. The ease of replication of polymers by hot embossing and injection moulding combined with the ability to thermally seal the channels has enabled the fabrication of disposable devices. Importantly, devices based on polymers are able to maintain the functionality of conventional substrates of silicon or silica. Heckele and Schomburg have reviewed the range of polymers which have been used in the molding of micro/nanofluidic systems [1]. In recent years, a new class of thermoplastic polymer, cyclic olefin copolymer (COC), has been synthesised with a unique range of optical, thermal and mechanical properties. COC copolymers have shown an optical transparency extending into the UV and deep-UV spectral ranges [2]. Another attribute of COC has been the ability to tailor it's thermal and mechanical properties by variation in the percentage of norbornene incorporated during initial polymerisation [2]. COC has been synthesized by the copolymerisation of a cycloolefin (typically norbornene) with an olefin (ethylene). Increase in the norbornene content has acted to stiffen the main chain and prevent crystallisation of the amorphous structure. An increase in concentration of norbornene has correlated with an increase in the glass transition temperature, T_g, [3] and a decrease in the ductility of the copolymer [4].

The fabrication of microfluidic chips in COC has been reported using both hot embossing [5,6] and injection molding [7] techniques. However, little data exists on the effect of comonomer content and embossing parameters on the fabrication of micropatterns in COC. This paper examines for the first time the effect of important variables on the embossing of microfluidic channels in standard grades of COC. These grades were selected as providing substrates with a wide range of T_g and mechanical properties. The results of experiments on the optimisation of embossing parameters have been applied in the fabrication of flow focusing devices for microdroplet generation.

EXPERIMENTAL DETAILS

The test pattern was comprised of an array of flow focusing devices containing channels of 50, 100 and 200 μm width. The masters for the devices were fabricated in dry film resist (Shipley 5038) as either a single (~35 μm) or a double (~70 μm) layer. The channels were lithographically patterned using a collimated UV source and transparency film masks with subsequent development in a 20% Na_2CO_3 solution. The resulting masters consisted of channels with smooth sidewalls and uniform width. The resist pattern was subsequently replicated as a Ni shim using an initial sputter deposition of 100 nm Ni followed by electroplating to a thickness of 150 μm. The Ni shims provided a tool for subsequent hot embossing of the COC substrates. Six grades of COC (Topas 9506, 8007, 5013, 6013, 6015 and 6017) were selected in order to provide a range of norbornene content and glass transition temperature, T_g as listed in Table I (Topas Datasheet). Slides (25 x 75 x 1.5 mm) of the 6 grades were supplied by Polyplastics Co., Tokyo. Also included in the experiments were slides of polymethylmethacrylate (PMMA) (Goodfellow). Initial experiments examined the depth of embossing versus force (6.75-22.5 kN) at a temperature 20 °C above T_g for 120 s. In a second series of experiments, the embossing time was varied (5-120 s) at a constant temperature 20 °C above T_g of each material and pressure (22.5 kN). Further experiments examined the depth of embossing versus temperature (30-180 °C) at a constant pressure (22.5 kN) and time (120 s). After de-embossing at room temperature, the depth of the patterns was measured at several locations along the channels by stylus profilometry. Scanning electron microscopy was used to examine the surface morphology of the channels.

Table I. Properties of Topas COC grades (Topas Datasheet)

COC Grade	**9506**	**8007**	**5013**	**6013**	**6015**	**6017**
Norbornene wt%	61	65	75	76	79	82
T_g (°C)	65	80	134	138	158	178
Heat Deflection Temp. at 0.45 MPa (°C)	60	75	127	130	150	170

RESULTS AND DISCUSSION

Measurements of Embossing Depth

An initial series of experiments was used to determine the embossed depth in 200 μm wide channels as a function of either the applied force (6.75-22.5 kN) or duration (5-120 s) at a temperature 20 °C above T_g of each material. In Figure 1(a), the embossed depth in various COC grades was shown to increase only slightly with force. The depth was similar for all grades of COC. Figure 1(b) shows that the depth remained approximately constant with variation in the duration of emboss at a temperature above T_g for times longer than 15 s. An emboss time of 120 s and force of 22.5 kN were selected in the subsequent experiments as sufficient to give a maximum depth without thermal distortion of the slides.

Figures 2(a) and (b) show measurements of the embossed depth versus temperature in the 6 grades of COC and in PMMA for 35 and 70 μm deep shims, respectively. Each point in these plots represents an average of 5 measurements at 1 cm spacing on a separate sample.

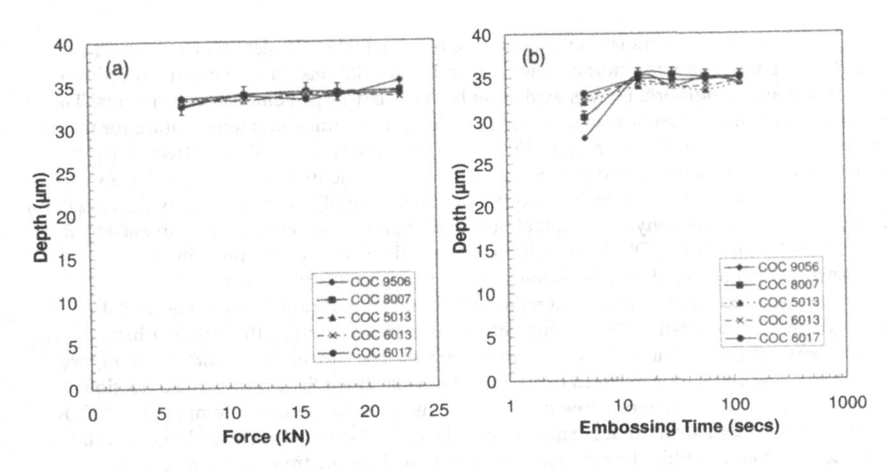

Figure 1. Embossing depth versus (a) force and b) time at a temperature ~20 °C above T_g.

Figures 2(a) and (b) show that, in all grades of COC and in PMMA, the embossed depth increased sharply over a critical range of temperatures. For PMMA, the transition temperature range was 100-120 °C, corresponding to it's T_g of 118 °C. Figure 2(a) shows that at temperatures below the critical range, the depth of embossing was 9-10 µm for all the COC grades and 20.7-21.9 µm in PMMA. At temperatures above the critical range, the embossed depth increased to a level of 31.2-35.1 µm and remained approximately constant. This depth of embossing was equal to the height of the ridge (35 µm) in the Ni shim. Similar trends in depth versus embossing temperature were evident in the deeper ~70 µm channels as shown in Figure 2(b). At temperatures below the transition in Figure 2(b), the embossed depth was 37.0-45.8 µm in all grades and ~51.9-52.3 µm in PMMA. Above the transition, the embossed depth of 69.6-70.8 µm was also equal to the height of the ridge in the shim (70.6 µm).

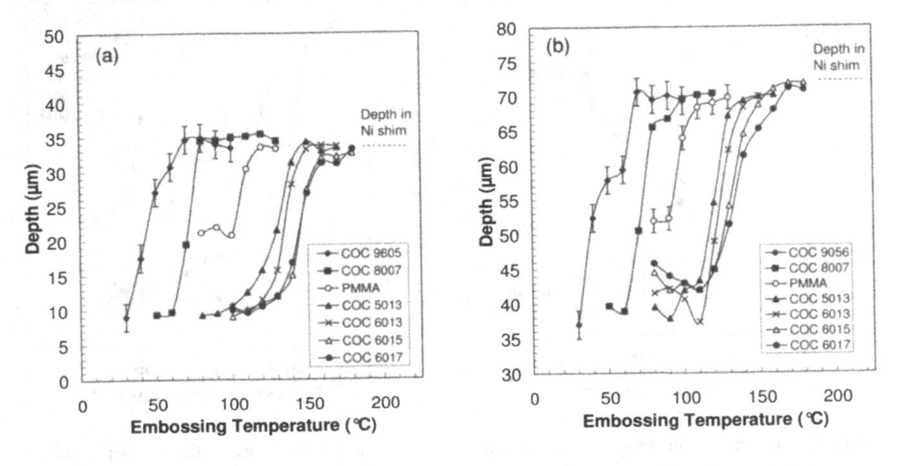

Figure 2. Embossed depth versus temperature in 200 µm wide channel measured using (a) ~35 µm and (b) ~70 µm deep shims. Applied force of 22.5 kN and embossing time of 120 s.

Figure 3(a) shows the transition temperature obtained from the depth measurements plotted as a function of wt % norbornene. A near linear relationship between the transition temperature and % norbornene was evident in both 35 and 70 μm embossed structures. The two plots were almost identical with a slight levelling off in transition temperature for the 35 μm deep structures. The linear trend in Figure 3(a) was consistent with the effect of rigid cyclic monomer units associated with norbornene in the structure. Forsyth et.al. have shown that the concentration of norbornene incorporated in the copolymer was directly dependent on the ratio of norbornene/ ethylene [3]. Increase in norbornene has correlated with enhanced hardness and stiffness in COC due to additional metallocene ring structures in the norbornene-ethylene copolymer backbone [8]. Above the transition, the range of temperatures tested corresponded to a region of viscous liquid flow in which deformation of the polymer was irreversible [9]. During deformation in this regime, the polymer behaved as a highly viscous liquid consistent with a non-compressible Newtonian liquid [9]. In Figures 2(a) and (b), the constant embossed depth for all grades above their respective transition temperatures was indicative of a low modulus regime of deformation. The modulus of COC has been previously shown to decrease sharply above T_g due to the increased ease of relative movement of chains within the structure [8]. In Figure 3(a), the transition temperature obtained by embossing experiments was consistently lower than the values of T_g measured by thermal analysis (Table 1). The occurrence of viscous deformation in an amorphous polymer has been previously reported to commence at a temperature significantly below T_g [10]. Hence, the transition temperature measured by the extent of deformation during embossing may consistently give a lower value than a transition temperature measured on the basis of structural change.

Figure 3(b) shows a plot of the average surface roughness, R_a, measured along the base of the 70 μm channels versus embossing temperature. Figure 3(b) shows that for all grades of COC, the magnitude of R_a decreased continuously with temperature until levelling-off. The minimum value of R_a ~10 nm at levelling off corresponded to the inherent roughness of the base plate used in initial lithographic patterning. The embossing temperature at levelling-off in R_a increased with T_g of the copolymer grade (cf. Figure 3(b) and Table I).

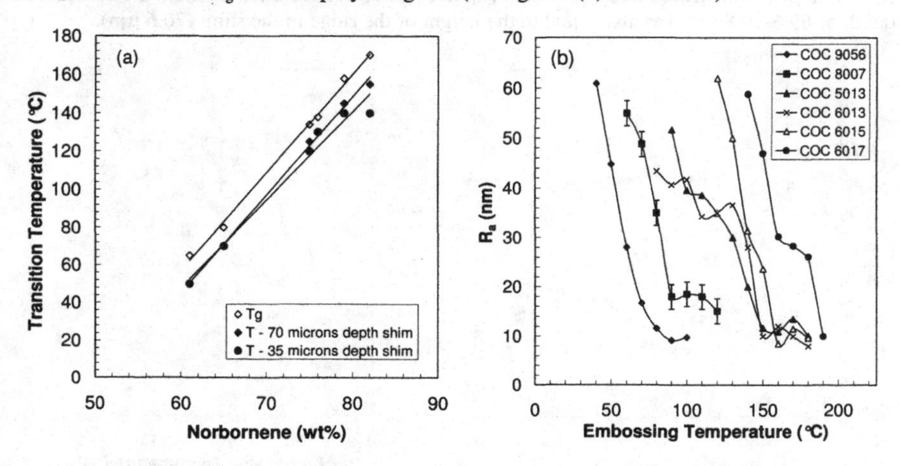

Figure 3. (a) Transition temperature (measured as the mid temperature in the range) versus norbornene content plotted for each grade of COC and (b) R_a versus embossing temperature for the grades of COC.

SEM images

Figure 4 has compared the surface morphology of the channels in which 8007, 5013 and 6015 grades were embossed at temperatures below and above the transition. These grades were selected as examples of a range of T_g. A profilometry trace with a height/width ratio of ~4.25 has been superimposed on the SEM image. The embossed channels in Figure 4 were formed using 70 µm deep shims to provide a maximum deformation.

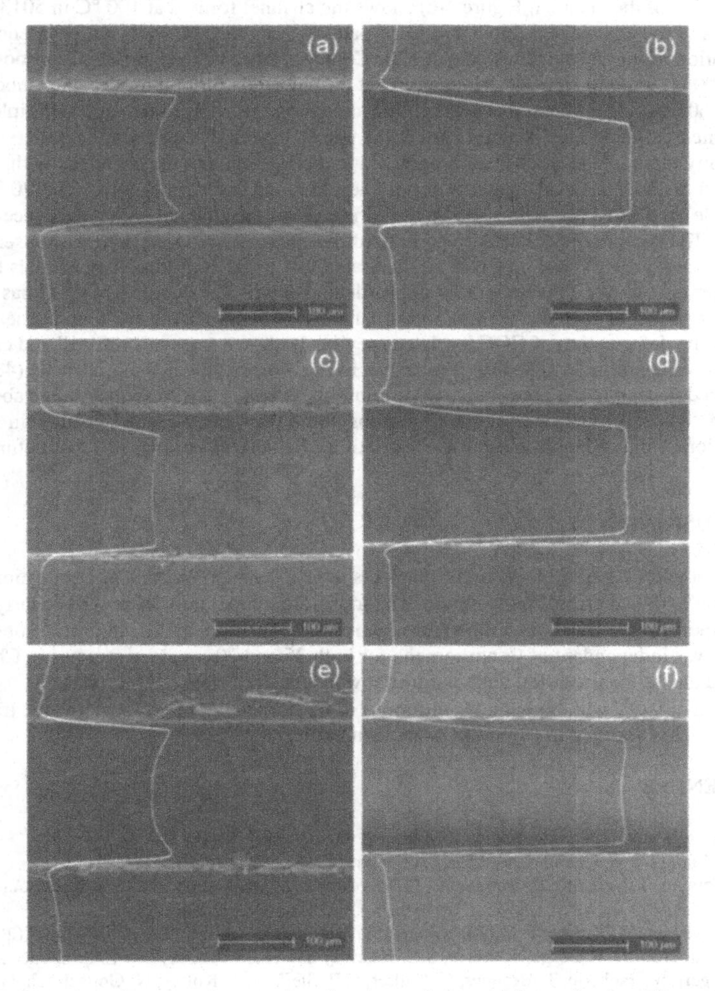

Figure 4. Scanning electron micrographs of embossed channel formed using 70 µm deep shims in 8007 grade at (a) 50°C and (b) 120°C; in 5013 grade at (c)100°C and (d) 170°C; and in 6013 grade at (e) 100°C and (f) 170°C. A profilometry trace (white line) has been imposed on the SEM image. The ratio of height/ width in the profilometry trace was 4.25.

Figure 4(a) shows that in the channel formed at 50 °C in 8007 grade, the edges of the sidewalls were exhibited a serrated pattern along the entire length with evidence of delamination. The profilometry trace has also shown a convex distortion at both the base of the channel and at the top surface. In Figure 4(b), a channel which was embossed in 8007 at a temperature of 120 °C (40 °C above T_g) has shown surfaces which were much smoother than at 50 °C with no signs of delamination at the sidewalls. A significantly greater depth of emboss occurred than at 50 °C. A slight ridge was evident in Figure 4(b) due to displacement on either side of the channel. Figure 4(c) shows the channel formed at 100 °C in 5013 grade. The edges of the channel exhibited a serrated pattern along the sidewalls with evidence of delamination. Figure 4(d) shows that, at 170 °C, the surfaces were significantly smoother than at 100 °C with no signs of delamination at the sidewalls. In Figure 4(e), the embossed channel (6013 grade embossed at 100 °C) has shown severe signs of damage. Multiple cracks have formed parallel to the channel both in the upper ridge and also along the base.

In many regions, the cracks had progressed to the sidewalls of the channels, with concurrent dislodgement of segments of polymer. In comparison, in Figure 4(f) (170 °C, 6013 grade), the SEM image has shown the formation of the channel with an absence of cracking. The smooth upper surface and base of the channel was associated with essentially featureless sidewalls. These results have illustrated that for all 3 grades, the channels formed at a temperature above T_g were similar in profile and depth. The ductility of COC has been previously shown to decrease with % norbornene [4]. At higher levels of % norbornene (68-78 wt%), the deformation of COC has been previously shown to occur as fibrillated crazes [4] which were similar in appearance to cracks in Fig. 4(e). A lower % norbornene (48 wt %) has resulted instead in the appearance of multiple shear bands [4]. At temperatures above the transition, the occurrence of a regime of viscous liquid flow has evidently resulted in deformation of the channels without the formation of a sheared or cracked substructure.

CONCLUSIONS

The hot embossing of microfluidic channels in COC was dominated by a transition in embossed depth at a critical temperature. The transition temperature increased linearly with % norbornene in the COC. At temperatures above the transition, the depth of the embossed channels was independent of % norbornene for both 35 and 70 μm deep structures. Channels embossed above the transition temperature showed a smooth sidewall morphology. Embossing of COC with higher concentrations of norbornene (≥75 wt%) below the transition temperature has resulted in extensive crack formation.

REFERENCES

1. M. Heckele and W.K. Schomburg, *J. Micromech. Microeng.* **14,** R1 (2004).
2. W.J. Huanga, F.C. Changa and P.P.J. Chu, *Polym.J.* **41,** 6095 (2000).
3. J.F. Forsyth, T. Scrivani, R. Benavente, C. Marestin and J.M. Perena, *J.Appl.Polymer Sci.* **82,** 2159 (2001).
4. V. Seydewitz, M. Krumova, G.H. Michler, J.Y. Park and S.C. Kim, *Polymer* **46,** 5608 (2005).
5. N.S. Cameron, H. Roberge, T. Veres, S.C. Jakeway and H.J. Crabtree, *Lab Chip* **6,** 936 (2006).
6. J. Steigert, S. Haeberle, T. Brenner, C. Müller, C.P. Steinert, P. Koltay, N. Gottschlich, H. Reinecke, J. Rühe, R. Zengerle and J. Ducré, *J.Microm.Microeng.* **17,** 333 (2007).
7. D.A. Mair, E. Geiger, A.P. Pisano, J.M.J. Fréchet and F. Svec, *Lab Chip* **6,** 1346 (2006).
8. T. Scrivani, R. Benaventure, E. Perez and J.M. Perena, *Macromol.Chem.Phys.* **202,** 2547 (2001).
9. L.J. Guo, *J.Phys D: Appl.Phys.* **37,** R123 (2004).
10. X. Shan, Y.C. Liu and Y.C. Lam, *Microsyst.Technol.* **14,** 1055 (2008).

Mater. Res. Soc. Symp. Proc. Vol. 1191 © 2009 Materials Research Society 1191-OO03-19

Biohybrid Photoelectrochemical Nanoengineered Interfaces

Arati Sridharan, Jit Muthuswamy, and Vincent B. Pizziconi

Harrington Department of Bioengineering, Arizona State University, MB 9709

Tempe, AZ 85287, U.S.A.

ABSTRACT

Incorporation of biophotonic components in artificial devices is an emerging trend in exploring biomimetic approaches for green technologies. In this study, highly efficient, nanoscaled light antenna structures from green photosynthetic bacteria, known as chlorosomes, comprised of bacteriochlorophyll-c pigment arrays that are stable in aqueous environments are studied in an electrochemical environment for their photoelectrogenic capacity. Biohybrid electrochemical cells containing chlorosomes coupled to the native bacterial photosynthetic apparatus have a higher dark charge storage density (at least 10-fold) than electrochemical cells with decoupled chlorosomes. Nevertheless, upon light stimulation, the charge storage density, also known as charge injection capacity, for both electrochemical systems increased the charge stored near the electrode. Decoupled chlorosome-based systems showed a light-intensity dose-dependent response, reaching a maximum change of ~300 nC/cm^2 at near sunlight intensities (~80-100mW/cm^2). Chronoamperometric studies under light stimulation conditions confirmed the photo-induced effect. Current studies are focused on optimization of the electrode/chlorosome interfacial properties across various heterogeneous interfaces. Successful implementation of harvesting photo-energy using the chlorosome or its derivatives may lead to substantial innovations in current biophotonic technologies, such as biofuel cells and retinal prosthetics.

INTRODUCTION

A key aspect of current lab-on-chip and biotechnology instruments occurs at the interface between the biological transducing element and the electrode. Interfacial energy transfer from a biomaterial to the artificial electrode system is difficult to characterize and often involves building the entire device to assess performance. In this study, a design parameter known as the charge storage density (CSD) is used to assess heterogeneous, biohybrid interfaces for potential optoelectronic devices. CSDs are typically used in the neuroscience and biomedical fields to assess the charge stored near the electrode/electrolyte interface in order to establish safe neurostimulation parameters [1,2]. As a novel extension of the concept, light-stimulated effects on photosynthetic materials derived from the green bacteria, *Chloroflexus aurantiacus*, are investigated in an electrochemical manner. In particular, the main light-harvesting structure of the green bacterium, known as the chlorosome, is studied for its photoenergy transfer capabilities in the absence of its native photosynthetic machinery. The chlorosome is a unique, nanoscaled structure, which is composed of arrays of self-assembled bacteriochlorophyll-c (BChl *c*) pigment molecules enclosed in a lipid monolayer [3]. Its extraordinary, natural properties, such as a 92% quantum efficiency, its relative independence from biological proteins for its light harvesting function, and high stability in aqueous solutions, make the chlorosome a worthy biophotonic candidate for potential use in novel optoelectronic devices [4,5]. Previous studies indicate that decoupled chlorosomes modulate the interfacial charge transfer properties at the electrode interface via the ground redox state of the BChl *c* pigments present within its body [6]. In this present study, chlorosomes embedded in bacterial fragments and purified chlorosomes were assessed and compared under various electrochemical conditions. In addition, the CSDs were compared to differences in the concentrations of oxidized to reduced pigments using absorbance spectroscopy.

EXPERIMENTAL DETAILS

J-10-Fl strain of *C.aurantiacus* bacteria were grown using previously described methods at of 300-500 lux at 55° C under anaerobic conditions in a modified Nitsch's D-medium [7]. Bacterial fragments containing chlorosomes were generated by ultrasonication of various concentrations of bacteria at 25 W for 5 minutes. Chlorosomes were purified using a similar protocol of Feick and Fuller [8]. Briefly, green-colored cultures were harvested by centrifuging at 3000 × g for 60 minutes after 7 days of growth (late exponential phase), re-suspended in 2 M sodium thiocyanate buffer (1:4 (w/v)), homogenized at 4 °C, and then French-pressed (20,000 psi) to obtain fragments containing chlorosomes. The resulting mixture was ultracentrifuged at 100,000 × g for 18 hours at 4 °C in a 5%-40% sucrose gradient. Chlorosome-containing fractions devoid of reaction centers were pooled, re-centrifuged in sucrose gradients, and dialyzed 6 times into 0.2 M phosphate buffered saline (PBS) (pH 8.0) to remove any extraneous pigments and smaller soluble contaminants from the photosynthetic apparatus. Absorbance spectroscopy at 470 nm, 740 nm and 795 nm was used, where the former two related to BChl *c* aggregates in the chlorosome and the latter to BChl *a* pigments present in the baseplate. The absence of reaction centers were confirmed by the lack of peaks at 805/866 nm. The number of chlorosomes was determined using empirical relationship derived by LaBelle [9].

A custom electrochemical cell with a graphite-based working electrode and platinum counter electrode were placed 1 cm apart. The main electrolyte used was 0.2 M phosphate buffered saline (PBS) at pH 8.0. A light-blocked silver/silver chloride reference in 3 M KCl (BASi, West Lafayette, IN) was placed equidistant from the other two electrodes. For typical, isolated chlorosome-based electrochemical studies, ~10^9- 10^{12} chlorosomes/ml with graphite as the working electrode was used. Electrochemical cells (working electrode side) were stimulated with a 150 W halogen lamp (400-850 nm spectrum). Using a IL-1700 research radiometer (International Light Inc., Newburyport, MA), the light intensities were kept constant at ~80 mW/cm^2 for most studies. For dose-dependent responses, the stimulating halogen lamp was mechanically modulated from ~30-80 mW/cm^2.

Electrochemical characterization was done using cyclic voltammetry using a CHI-660a potentiostat (CH Instruments, Austin, TX). CSDs are calculated as the average area under the cyclic –voltammetric curve to attain a maximum limit of charge injection in an electrochemical environment. Cyclic voltammetry under light and dark conditions was performed at 50 mV/s between -0.2 to 0.8 V for isolated chlorosomes, and -0.3 to 0.9 V for solutions containing bacterial fragments. Charge storage density was calculated as the average area between the two sweeps of the CV curves, which was determined using numerical integration techniques [1]. The calculated total charge was then divided by the effective electrode surface area to obtain a charge storage density value. The typical working electrode surface area was estimated using the Cottrell relationship using a similar protocol to Cummings et al [10]. Briefly, oxidative current at the electrodes was measured with 1mM ferricyanide in 1M KCl in aqueous solution (diffusion coefficient ~ 6.3×10^{-5} $cm^2 s^{-1}$) under chronoamperometric conditions for 20 seconds at 0.7 V with 2 seconds quiet time at 0 V. The effective electrode area for 3 samples was determined to be 0.67 ± 0.09 cm^2 from the calculated slope of $I = 1/\sqrt{t}$ where I represent current, and t represents time in the Cottrell equation.

DISCUSSION

Figure 1a shows a typical cyclic voltammogram (CV) of an electrochemical cell containing chlorosomes coupled to the native photosynthetic machinery within bacterial fragments under dark conditions. Compared to an electrochemical cell with only PBS electrolyte, the CV for the bacterial fragments generates more current at higher voltages as compared to PBS controls. This suggests that at higher voltages, more elements within the electrochemical cell containing bacterial fragments get oxidized at the electrode interface. Additionally, the larger width of the CV curve for bacterial fragment containing cell suggests that the electrode is more capacitative in nature. The general increase in capacitance is due to the highly complex structures present within the cell where chlorosome containing bacterial fragments have a milieu of organic, pigment aggregates, proteins, lipids, and other biological materials.

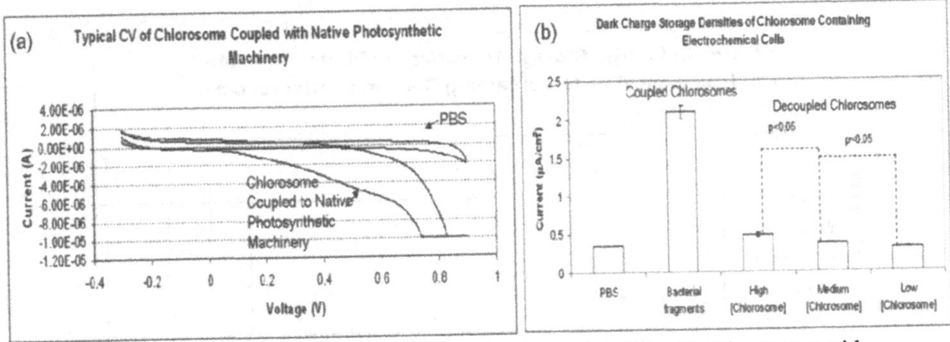

Figure 1. (a) A typical cyclic voltammogram of an electrochemical cell with chlorosomes with entire photosynthetic machinery compared to that with only PBS electrolyte. (b) Comparative dark CSDs of various electrochemical cells containing chlorosomes under dark conditions.

For a more quantitative analysis, CSDs for various electrochemical cells in the dark were calculated and compared in figure 1b. Calculated as the average area uncer the CV curve, the CSD parameter would represent the total charge that could be stored at the electrode interface. An electrochemical cell containing an extremely complex pigment-protein structure like the chlorosome coupled to the photosynthetic machinery had ~10 times the interfacial charge density compared to PBS controls. Interestingly, in electrochemical cells containing only decoupled chlorosomes (i.e in the absence of the native photosynthetic machinery), the CSD parameter decreased to near PBS control levels. This suggested that the capacitative nature of the electrochemical cell as seen in electrochemical cells with coupled chlorosomes was removed, indicating a minimal presence of high dielectric materials such as lipids and proteins. Additionally, the nanoscale size of the decoupled chlorosome particles (~100 nm × 30 nm × 10 nm) did not seem to hinder the relative interfacial charge transfer properties.

However, high and low concentration differences between electrochemical cells containing only decoupled chlorosomes were found to be significant. This indicated that high decoupled chlorosome concentrations (~10^{12} chlorosomes/ml) had an incrementally higher CSD compared to medium (~10^{10} chlorosomes/ml) and low (10^{12} chlorosomes/ml) concentrations. Since the nanoscaled structure of decoupled chlorosomes is known to be composed mainly of highly-charged, pigment aggregates, the increase in the relative charge stored near the electrode would be plausible.

Figure 2 shows the changes in the calculated CSD parameter relative to various white light intensities of electrochemical cells containing decoupled chlorosomes. As seen in the figure, as light intensity increases, the charge stored near the electrode increases by ~300 nC/cm^2 at near sunlight level intensities (~80-100 mW/cm^2). The white light intensities were radiometrically measured from a tungsten filament source using an IL-1700 research radiometer. At lower light intensities (~30 mW/cm^2), the relative change in charge density was ~1/10 (~30 nC/cm^2) of the high intensity measurement for the same concentration of chlorosomes. This would suggest that

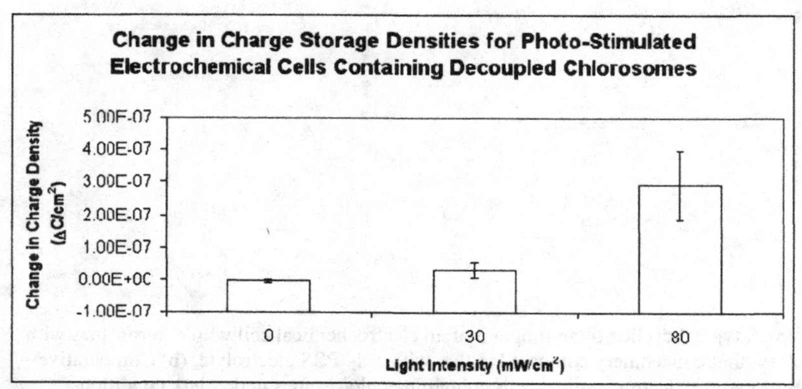

Figure 2. Photo-stimulation effects on CSDs of electrochemical cells containing decoupled, isolated chlorosomes at various light intensities.

logarithmic changes in light-induced charge density occurred at the electrode interface. Additionally, at high light intensities, the relative light-induced change in the CSD parameter for electrochemical cells with coupled chlorosomes was approximately half (~120 nC/cm^2) of the decoupled chlorosome electrochemical cells despite a higher capacitance and chlorosome concentration in these systems. Even though coupled chlorosome systems had a higher dark charge storage density compared to decoupled systems, the relative photochange in charge was much lower. This could be due to extraneous recombination effects from the presence of various reactive species represent within the chlorosome-containing bacterial fragments. The removal of these extraneous species via dialysis and other purification methods in generating decoupled chlorosomes may have facilitated and relatively stabilized the increase of photo-induced charges stored at the interface.

Mechanistically, the BChl *c* pigment-aggregates within the chlorosome body were expected to play a role in the increase in the CSD parameter. In the natural system, the BChl *c* pigments would be found in the reduced state. In a separate experiment using absorbance spectroscopy, the relative differences between photo-oxidized and reduced pigment concentrations were found to correlate with the relative increase in the light-induced CSD parameter [11]. This implied that light stimulation increased the charge stored near the electrode interface due to a photo-oxidation mechanism. Additional experiments would be necessary to investigate the stoichiometry and kinetics of charge transfer at the electrode interface in a photoelectrochemical cell.

CONCLUSIONS

This study shows a novel method to characterize the optoelectronic, biohybrid interface using the charge storage density (CSD) parameter. In this study, a complex photosynthetic element composed of highly-organized, self-assembled pigment arrays known as the chlorosome, was found to increase in the relative photo-induced change in charge densities stored near the electrode interface. The CSD parameter was also used to analyze relative concentration differences in various electrochemical cells containing chlorosomes. It was observed that incremental changes in chlorosome-particle concentrations increased the CSD. The concept of the CSD parameter, which is an extension from neuro- and tissue stimulation studies, may be utilized in a broader sense for characterizing relative charge density changes at the heterogeneous, bio-electrode interface.

ACKNOWLEDGMENTS

We acknowledge the National Science Foundation (NSF) funded Integrated Graduate Education and Research Training (IGERT) grant in Optical Biomolecular Devices: From Natural Paradigms to Practical Applications (NSF0114434) for generous graduate support (PI: Dr. Neal Woodbury, Department of Chemistry & Biochemistry, Biodesign Insitute, ASU). This study was also partly funded by the Graduate and Professional Students Association (GPSA) dissertation grant (2005-2006) and the generous support of the Faculty Emeriti Association at Arizona State University (ASU).

REFERENCES

1. E. W. Keefer, B .R. Botterman, M. I. Romero, A. F. Rossi, G. W. Gross, *Nat. Nanotechnol.* **3,** 434 (2008)
2. R. Saha and J. Muthuswamy, *Biomed. Microdevices* **9,** 345 (2007).

3. R. E. Blankenship, *Molecular Mechanisms of Photosynthesis*, (Blackwell Science, 2002), pp. 63-81.
4. J. T. LaBelle, Dissertation, Arizona State University (2001).
5. J. M. Olson, *Photochem. Photobiol.* **67**, 61 (1998).
6. A. Sridharan, J. Muthuswamy, J. T. LaBelle, V. B. Pizziconi, *Langmuir* **24,** 8078 (2008).
7. P. D. Gerola, J. M. Olson, *Biochim. Biophys. Acta* **848,** 69 (1986).
8. R. G. Feick, R. C. Fuller, *Biochemistry* **23,** 3693 (1984).
9. J. T. LaBelle, V. B. Pizziconi, U.S Patent 7067293, (2006).
10. E. A. Cummings, P. Mailley, S. Linquette-Mailley, B. R. Eggins, E.T McAdams, S. McFadden, *The Analyst* **123,** 1975 (1998).
11. A. Sridharan, J. Muthuswamy, V. B. Pizziconi, *Langmuir*, (2009), (in press).

Materials Synthesis on Chip

Mater. Res. Soc. Symp. Proc. Vol. 1191 © 2009 Materials Research Society 1191-OO04-10

Microfluidic Channel Fabrication in Dry Film Resist for Droplet Generation

Patrick W. Leech[1], Nan Wu[2] and Yonggang Zhu[2]
[1]CSIRO Materials Science and Engineering, Clayton, 3168, Victoria, Australia
[2]CSIRO Materials Science and Engineering, Highett, 3190, Victoria, Australia

ABSTRACT

Dry film resist has been used in the fabrication of Masters in microfluidic devices for droplet generation. The minimum feature size in the resist was controlled by the type of mask (transparency or electron beam Cr mask), the resolution of the pattern in transparency masks (2400 or 5080 dpi) and thickness of resist in the range from 35 to 140 μm. The Master patterns formed in dry resist were replicated as a Ni shim and then hot embossed into Plexiglas 99524. These devices were used to generate water-in-oil droplets with a well defined dependence of diameter and frequency on flow parameters. The application of dry laminar resist and transparency masks has allowed the rapid fabrication of prototype devices.

INTRODUCTION

Microfluidics systems have in recent years been used as a tool for the generation of microdroplets [1]. In particular, the incorporation of a flow focusing orifice in microfluidics systems has provided a means of forming a uniform, continuous stream of droplets of a specified size [1]. The function of the orifice has been to create an instability in flow within two immiscible fluids by hydrodynamic focusing. The selection of the channel widths and design of the flow focusing nozzle in these structures has allowed the control of droplet size within a narrow range (~3%) at frequencies of ≥1 kHz [1,2]. A promising application of such highly uniform droplets has been in the synthesis of solid microparticles containing a controlled structure of layers [3]. Droplets with an outer shell structure of this type have also been used to enclose a suspension of micro- or nanoparticles [4]. Furthermore, the incorporation of minute quantities of DNA, protein or biological cells in microdroplets has provided a large number of separate reactor compartments for experimentation. These characteristics have recently been applied by Tawfig and Griffiths at Cambridge, UK in an investigation of the concept of in-vitro compartmentalisation (IVC) [5].

The flow focusing devices used in droplet generation have predominantly been constructed in polydimethylsiloxane (PDMS) (for example, [3, 6-8]). The fabrication of the master pattern for the molding of channel structures in PDMS has typically been based on lithographic definition in a layer of SU8 resist. In this paper, we have examined the alternate application of dry film resist in fabrication of the Master pattern. Intrinsic properties of dry film resists have included low cost, excellent adhesion to a range of substrates, no edge bead, fast prototyping and easy removal. Dry film resists have been most commonly used in the patterning of printed circuit boards (PCBs) but have also recently been applied as moulds for electroplating in a LIGA type process [9] and soft lithography [10] and as structural elements in microfluidic chips [11,12]. In this paper, we have examined the first application of dry film resist in the fabrication of microfluidics devices for droplet generation. Dry laminar resist and high resolution transparency masks have been combined as a means of rapid prototyping of the Master patterns.

EXPERIMENTAL DETAILS

Shipley 5038 (negative film resist) of ~35 μm thickness was laminated as either a single layer (~35 μm) or multiple layers at 113 °C onto a plate of polished stainless steel. The resist was then lithographically patterned using either a transparency mask (2400 or 5080 dpi) or an equivalent pattern in a Cr mask prepared by e-beam lithography. The masks contained test arrays of circles, squares and lines with a feature width/ spacing of 20-200 μm. After exposure using a collimated UV source (λ = 350-450 nm) at 16.5 W/cm^2, the pattern was developed in a solution of 20% Na_2CO_3. The patterned features were evaluated as a function of exposure dose (15-600 mJ/cm^2), the thickness of resist (35-140 μm) and the type of mask. Several designs of flow focusing chip were fabricated with channel widths of 50 μm - 200 μm. The patterned resist was replicated as a Ni shim using an initial sputter deposition of 100 nm Ni followed by electroplating to 150 μm thick. The Ni shims provided a tool for subsequent hot embossing of Plexiglas 99524 (Degussa) (PMMA) substrates (75 x 25 x 1.5 mm). A temperature of 150 °C and force of 25 kN was used in a press with flat upper and lower brass plates. A capping layer of Plexiglas 99524 was then thermally sealed onto the embossed surface of the chip to create capillary channels. The experiments in droplet generation were performed using de-ionised water and mineral oil (Sigma). The kinematic viscosity of the oil and water phases was 15.4 mPaS and 8.4 mPaS at 20°C with densities of 840 kg/m^3 and 1000 kg/m^3, respectively. Interfacial tension between the oil and water phases was ~0.06 N/m. Both the water and oil streams were generated by neMESYS equipped with SGE syringes (250 mL capacity). The syringes had been fitted with high grade filters in order to eliminate any microscopic contaminant in either phase. The size of droplets was measured by analysis of frames in a video camera recorded at 200-250 frames/s.

RESULTS AND DISCUSSION

Characteristics of 5038 Resist

An initial series of experiments was performed in order to define the resolution limits of the resist as a function of processing parameters. Fig. 1(a) shows the response curve for a single layer of resist. The developed film thickness was measured as the maximum height of the ridge which was patterned at progressively higher doses. The dose required for the commencement of cross linking, D_g^i, was measured as ~16.5 mJ/cm^2 while the dose for 100 % polymerisation, D_g^o, was identified as ~60 mJ/cm^2. The contrast, γ, of the 5038 resist was determined as $\gamma = 1/[\log D_g^o - \log D_g^i] = 1.66$, similar to the maximum value of γ in SU8 resist (γ = 1.7) after a post- bake at 100 °C.

The transparency masks consisted of a dot matrix of printed features in black ink on a transparent Mylar film. Dot size in the 2400 and 5080 dpi masks was 10.5 μm and 5 μm, respectively. The equivalent Cr mask comprised a conventional etched pattern on quartz. These types of masks correlated with a range of cost from ~$15 for a 2400 dpi transparency, ~$100 for a 5080 dpi transparency and ~$800 for Cr masks.

The minimum feature size, δ, was determined as the smallest width dimension/ spacing which was reproducibly fabricated in resist. Figure 1(b) shows a plot of δ above the critical dose versus thickness of resist for the different types of mask. A curve of best fit for the data points in Figure 1(b) indicates an exponential dependence of δ on resist thickness. For a single layer of resist (35 μm thick) using 2400 dpi transparency masks, the results have shown that the minimum feature size for these equal size/ gap spacing features was ~60 μm. The use of a 5800 dpi transparency mask increased the attainable resolution for patterns to

30-40 µm. In comparison, the application of a Cr mask has enabled the fabrication of circles and squares with a 25 µm dimension in a single resist layer. With 2 layers of laminated resist (70 µm thick), the minimum feature size was increased to 100 µm for 2400 dpi and 50-60 µm for 5080 dpi film masks. Increase in dot size in the transparency masks has evidently resulted in a more irregular edge with a reduced ability to pattern finer structures. δ was also limited by enhanced light scattering with increasing thickness of the resist. Use of the Cr mask has enabled the patterning of features of 35 µm width/ spacing at 70 µm thickness, equivalent to an aspect ratio of 2:1. Previously, Vulto et.al. [14] have reported a minimum feature size of 20 µm in 54 µm thick laminar resist using Cr masks. Similarly, Stephan et.al. [12] have demonstrated features of 30-35 µm in a 35 µm thick layer of dry film resist with Cr masks.

Figure 2 shows an example of the channel structure in 5038 resist fabricated using a 5080 dpi transparency mask. In Figure 2(a), the channel shows smooth sidewalls at a depth of 70 ± 0.1 µm. Irrespective of the depth of resist (35 or 70 µm), the sidewall angle was measured by stylus profilometry as 75 ± 1°. The fine features on the sidewalls evident in Figure 2(a) were associated with the dot structure in the mask.

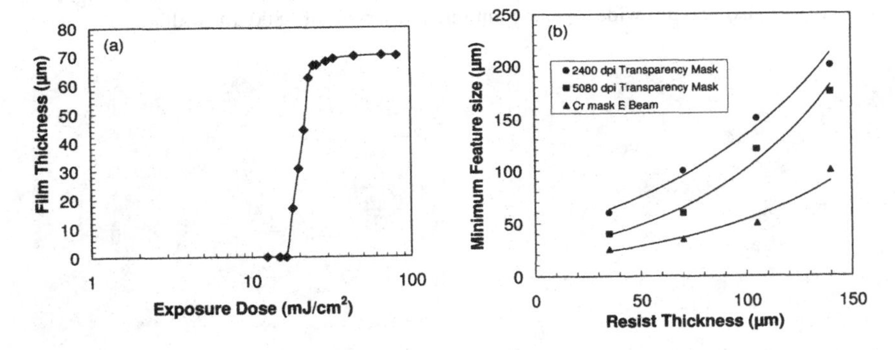

Figure 1. a) Response curve for a single layer of 5038 dry film photoresist and b) minimum feature size, δ, versus resist thickness using different types of mask.

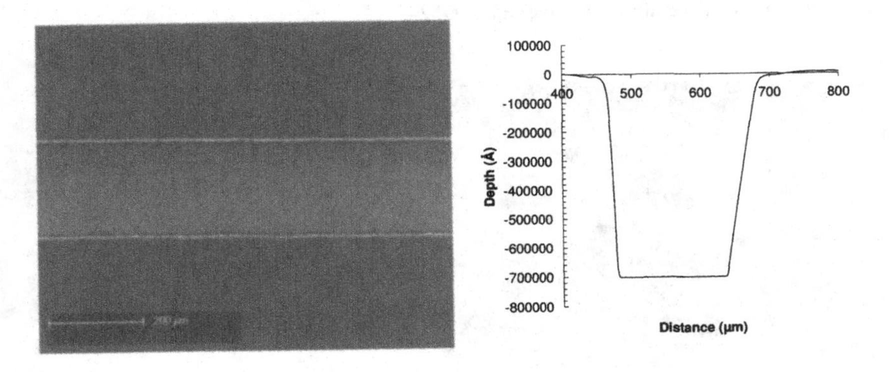

Figure 2. (a) 70 µm depth channels patterned in 5038 resist using a 5080 dpi transparency mask. (b) a profilometry trace of the channel.

Application of 5038 resist in fabrication of droplet generation chips

Master patterns of several types of flow focusing chip were fabricated in 5038 resist using 5080 dpi transparency masks. Figure 3 compares the features of the flow focusing junction in the resist Master with a corresponding embossed polymer. The master pattern in Figure 3(a) has shown clear definition of features including the 50 μm channels. The irregularities on the sidewalls due to the mask were visible. Replication of the 70 μm deep structure has produced relatively smooth surfaces along the channels. A rounding of the upper edges originated in the Ni shim. Fig. 4(a) indicates the direction of fluid flow in the orifice region in a droplet chip which was embossed in Plexiglas. A stream of water-in-oil droplets was generated by pumping an aqueous solution through the left-hand side of the converging channel while the oil solution was pumped from the two vertical inlets. The symmetric oil branches and the water branch intersected at the inlet of the production channel which was directed towards downstream for the processing or waste. Droplets were formed when the aqueous jet became fragmented due to the shear force exerted by the oil streams. The vertical channels were 200 μm wide, reducing to ~50 μm at the inlet of the junction. The horizontal outlet on the right hand side was also ~50 μm wide, opening out into a reservoir of ~800 μm width.

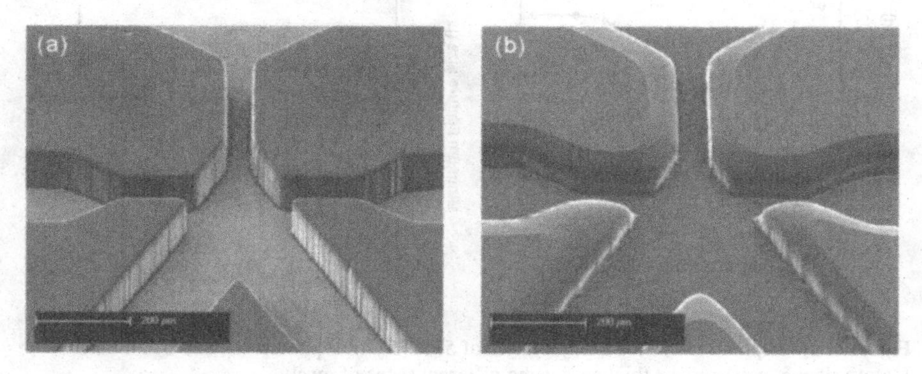

Figure 3. (a) Orifice region patterned in dry film resist (70 μm thick) using a 5080 dpi transparency mask, (b) embossed in Plexiglas 99524.

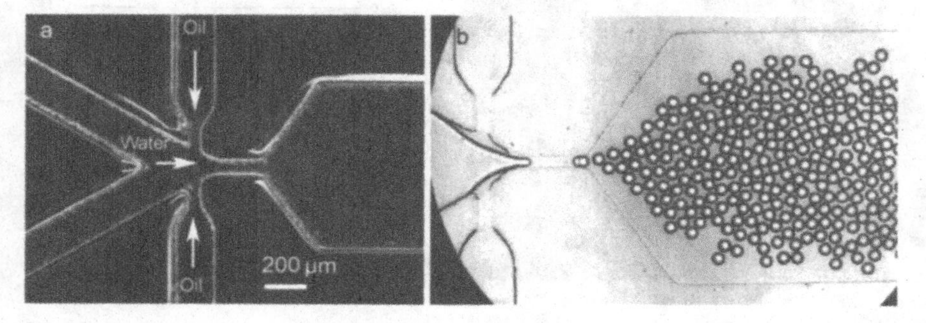

Figure 4. (a) Embossed flow focused pattern in Plexiglas 99524 showing direction of liquid flow and b) sealed device used in the generation of water droplets (50.7 μm) in oil.

After sealing, the device shown in Figure 4(b) has produced individual droplets of a uniform diameter (≤3% variance in ~50.7 μm). The rate of droplet formation was controlled by the relative flow rates of the oil and water streams. In general, the frequency of droplet formation increased in the range from 20 to 100 droplets/sec with rise in the flow rate of water from 10 to 80 μL/h. At aqueous flow rates of ≤30 μL/hr, the rate of droplet generation was independent of the flow rate of oil. In contrast, at aqueous flow rates of ≥60 μL/hr, the increase in droplet frequency with aqueous flow rate was greater for higher oil flow rates. Hence, at higher aqueous flow rates, an increase in the rate of droplet formation occurred with an increase in oil flow rate.

Droplets containing green-fluorescent protein (GFP) (asFP499) were generated in a Plexiglas flow focusing system with one aqueous inlet and detected using laser induced fluorescence (LIF). Figure 5 has compared the fluorescence intensity obtained by counting 500 droplets at each concentration subtracted for background noise. In these experiments, the oil and water flow rates were 20 μL/h and 10 μL/h, respectively. As illustrated in Figure 5, each spike represented one droplet passing through the detection point with excitation by the laser induced fluorescence at 488 nm with a bin time of 10 ms. The concentration of GFP solution was varied by dilution at 100, 500, and 1000 times using a phosphate buffer solution (PBS). As part of the detection procedure, a sample at each concentration was collected and calibrated in order to establish that higher concentrations of GFP droplets corresponded to a greater intensity of fluorescence. The results in Figure 5 indicate that each of the generated droplets contained a controlled concentration of GFP, as determined by the dilution. A similar level of background fluorescence was evident in the PBS buffer solution and the diluted samples. Background fluorescence has been previously linked to emission from the polymeric surface of the chip [13]. In the data shown in Figure 5, a filter block (FITC, Nikon) involving excitation and emission filters and a dichroic mirror was used to selectively excite fluorescence and to collect the emission with minimum optical background. A similar level of background was obtained at the different dilutions, indicating that the effect in Figure 5 was due to the ability to generate droplets with well defined characteristics.

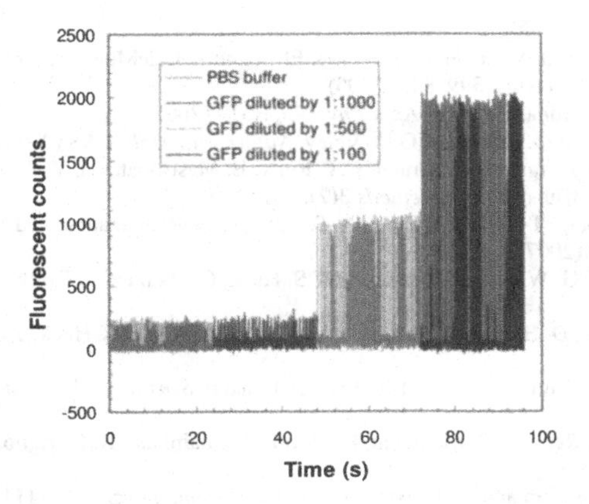

Figure 5. Fluorescence intensity of droplets which were generated at different dilutions of green fluorescent protein (GFP). The upper and lower markers at each concentration represented the fluorescence intensity of 500 droplets and the background level, respectively.

The combined use of dry film resist and transparency masks has enabled the fabrication of prototype Masters from the concept to patterning stages in less than 2 days. The fabrication process for the Master (resist lamination and lithographic exposure) is relatively simple with a total time of ~15 min. These short processing times and other intrinsic properties of dry film resist including uniform thickness distribution, low cost and excellent adhesion to a range of substrates mean that these are promising materials in the fabrication of microfluidic structures. The issue of limited resolution in dry resists has been addressed to some extent in recent years. For example, Zhao et.al have reported the fabrication of minimum line widths of 10 μm using 15 μm thick dry film resist by a Direct Projection Exposure technique [14]. However, to obtain structures with a higher aspect ratio (≥5), the application of alternate techniques such as patterning in SU8 resist or deep reactive ion etching is necessary.

CONCLUSIONS

Dry film resist has been applied in the fabrication of master patterns for microfluidic flow focusing systems used in droplet generation. The combination of a 70 μm thick layer of resist with a 5080 dpi film transparency mask has produced features with a resolution of 50 μm and an aspect ratio of 1.4. An aspect ratio of 2:1 at 70 μm thickness was obtained with Cr masks. The minimum feature size was shown to decrease with increasing resolution of the mask and decreasing thickness of resist. Masters which were fabricated in dry resist were replicated as a Ni shim and hot embossed in Plexiglas. The resulting devices have generated well defined droplets with variance in diameter of ≤3% in a single chip. The technique using dry film resist has advantages of rapid prototype fabrication, low cost and simplicity of process.

ACKNOWLEDGEMENTS

Ni shims were produced by F. Glenn and B. Sexton.

REFERENCES

1. Huebner, S. Sharma, M. Srisa-Art, F. Hollfelder, J.B. Ebel, and A.J. deMello, *Lab Chip* **8**, 1244 (2008).
2. M. Joanicot and A. Ajdari, *Science* **309**, 887 (2005).
3. G.F. Christopher and S.L.Anna, *J.Phys D:Appl.Phys*. **40**, R319 (2007).
4. J.R. Millman, K.H. Bhatt, B.G. Prevo and O.D. Velev, *Nature Materials* **4**, 98 (2004).
5. O.J. Miller, K. Bernath, J.J. Agresti, G. Amitai, B.T. Kelly, E. Mastrobattista, V. Tally, S. Magdassi, D.S. Tawfik and A.D. Griffiths, *Nature Methods* **3(7)**, 561 (2006).
6. W-L. Ong, J. Hua, B. Zhang, T-Y. Teo, J. Zhuo, N-T. Nguyen, N. Ranganathan and L. Yobas, *Sensors and Actuators* **A138** 203 (2007).
7. F. Courtois, L.F. Olguin, G. Whyte, D. Bratton, W.T.S. Huck, C.Abell and F.Hollfelder, *ChemBiochem* **9**, 439 (2008).
8. L.M. Fidalgo, G. Whyte, D. Bratton, C.F. Kaminski, C. Abell and W.T.S. Huck, *Angew.Chem.Int.Ed.* **47**, 2042 (2008).
9. H. Lorenz, L. Paratte, R. Luthier, N.F. de Rooij and P. Renaud, *Sensors and Actuators* **A53**, 364 (1996).
10. K. Stephan, P. Pittet, L. Renaud, P. Kleimann, P. Morin, K. Ouaini and R. Ferrigno, *J. Micromech. Microeng*. **17** N69 (2007).
11. Y-C. Tsai, H-P. Jen, K-W. Lin and Y-Z. Hsieh, *Journal of Chromatography A* **1111**, 267 (2006).
12. P. Vulto, N. Glade, L. Altomare, J. Bablet, L. Del Tin, G. Medoro, I. Chartier, N. Manaresi, M. Tartagni and R. Guerrieri, *Lab Chip* **5** 158 (2005).
13. A. Piruska, I. Nikcevik, S.H. Lee, C. Ahn, W.R. Heinman, P.A. Limbach and C.J. Seliskar, *Lab Chip*. **5** 1348 (2005).
14. S. Zhao, H. Cong and T. Pan *Lab Chip*. **DOI:** 10.1039/b817925e (2009).

Cell Manipulation and Biomimetics on Chip

Mater. Res. Soc. Symp. Proc. Vol. 1191 © 2009 Materials Research Society 1191-OO05-08

The Normal and Shear Strength of the Cell-Implant Interface: Accelerated Negative Buoyancy as a Method of Cell Adhesion Assessment

Helen J. Griffiths[1], C. Andrew Collier[1], T. William Clyne[1]
[1]Department of Materials Science and Metallurgy, University of Cambridge, Pembroke St, Cambridge, CB2 3QZ, UK

ABSTRACT

The strength of adhesion at the cell-substrate interface is an important parameter in the design of many prosthetic implant material surfaces, due to the desire to create and maintain a strong implant-tissue bond. This study focuses on the mechanical strength of the interface and the ease of cell removal from ceramic coatings using normal and shear forces, but also looks at cell proliferation rates on the same series of surfaces.

This systematic study of cell proliferation and adhesion has been carried out on a series of oxide coated Ti6Al4V-based substrates with a range of surface morphologies and chemistries. Oxide coatings were formed using Plasma Electrolytic Oxidation (the PEO process).

Cells were seeded at a low concentration onto substrates and proliferation monitored for up to three weeks. The same cell concentrations were seeded on samples for adhesion testing. These were cultured for a few days to ensure well established adhesion of viable cells. The normal and shear strength of osteoblasts (bone cells) and chondrocytes (cartilage cells) adhered to these substrates was measured using accelerated negative buoyancy within an ultracentrifuge.

The variation in proliferation rates on, and adhesive strengths to, the range of coatings, is discussed and related to morphological and chemical differences in the coatings. A comparison is made between the normal and shear strengths of the cell-coating bonds and the differences between the behaviour of the two cell types discussed.

INTRODUCTION

Implant tissue adhesion is a common requirement in for prosthetic implants. One method of improving adhesion is to apply a coating to an underlying prosthesis, hence maintaining the load bearing properties of an implant whilst increasing its security. There are a range of materials and coatings currently used for implants [1]. It has been suggested that coatings formed by plasma electrolytic oxidation are good candidates as coatings for implants [2-9].

Plasma Electrolytic Oxidation (PEO), generates well-adhered, wear-resistant oxide coatings with fine-scale interconnected porosity [10], as well as coarser porosity and surface roughness. Coatings are formed in an aqueous electrolyte, via the application of a potential of the order of a few 100 V, which causes a series of localised discharge events. Good adhesion to the underlying substrate minimises the worry of spallation in situ and the porosity and roughness are thought to be likely to improve cell and tissue adhesion.

Another advantage of PEO coatings is ability to tailor the coating chemistry and structure whilst by varying processing conditions. This study compares the normal and shear strength of the cell-substrate bond when bovine chondrocytes are adhered to 3 different forms of PEO on Ti6Al4V and to uncoated Ti6Al4V. An ultracentrifuge was used to apply stress to detach cells rather than more conventional methods such as the parallel plate flow chamber due to complications with such methods when a substrate is not perfectly flat [11].

EXPERIMENT

Substrate preparation and characterization

Samples of Ti6Al4V, in the form of discs of diameter 12.5 mm, were polished with 1200 SiC grit and, in some cases, then PEO coated, using either "phosphate", PEO(P); "mixed", PEO(M); or "aluminate", PEO(A) electrolytes. Surfaces were cleaned for 30 min in an ultrasonic bath of distilled water, followed by 30 min in acetone, and sterilized at 170 °C for 3 hours. Surface roughness was measured by optical profilometry and phases identified by X-ray diffraction. Ti6Al4V has a native surface layer of rutile, PEO(P) comprises a mixture of rutile and anatase, whilst PEO(M) and PEO(A) contain rutile, anatase and aluminium titanate. Microstructures are shown in Figure 1.

Cell Culture

Bovine chondrocytes were extracted from the metacarpophalangeal joints of young animals and cultured in Dulbeccos Modified Eagles Medium, DMEM, (Sigma) supplemented with 10% Foetal Bovine Serum (Sigma) and 1 % Penicillin-Streptomycin (Sigma). Human osteoblasts (ECACC) were cultured in McCoys 5A medium (Invitrogen) supplemented 10% Foetal Bovine Serum (Sigma), 0.1% Penicillin-Streptomycin-Glutamine (Invitrogen) and 30 $\mu g\ mL^{-1}$ vitamin C (Wako). Substrates were placed in 24-well plates and cells seeded at 10^4 cells per sample. Cells were cultured for 2 days prior to mechanical adhesion experiments. The proportion of cells remaining adhered to samples was estimated using the alamarBlue™ Assay [12] before and after the stresses were applied, assuming the degree of fluorescence to be approximately proportional to the number of metabolising cells.

Adhesion Assays

Normal and shear forces were both created by accelerated buoyancy within a swing-bucket ultracentrifuge. To apply normal forces substrates were inverted, with adherent cells, within centrifuge tubes of culture medium. To apply shear forces samples were held within similar tubes parallel to the tube axis. A range of forces was applied by adjusting the centrifuge rotation frequency between runs. Six repeats for each sample type at each spin frequency were performed.

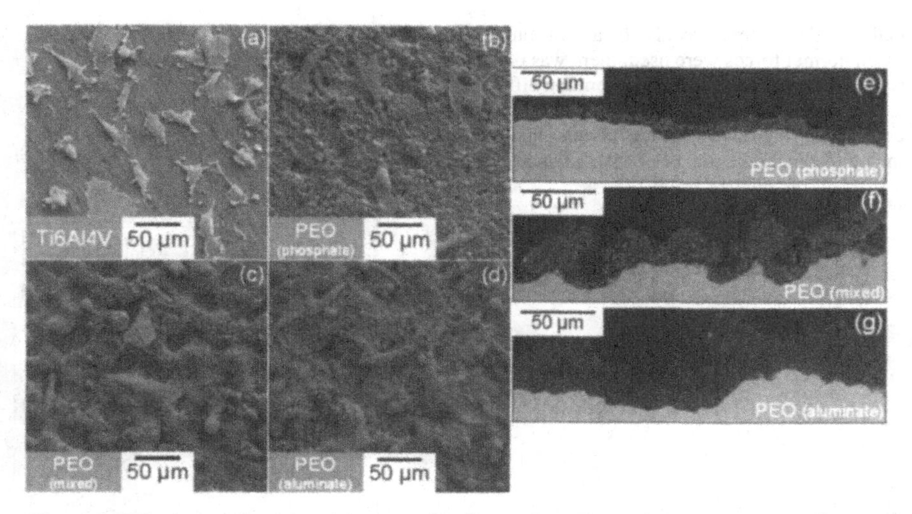

Figure 1: SEM micrographs of the substrates used (a-d) secondary electron images of cells adhered to samples, cells coloured green (e-g) back scattered electron images of cross sections through the coatings.

RESULTS AND DISCUSSION

Normal and shear forces were applied to cells adhered to substrates via negative buoyancy within an ultracentrifuge. The ease of removal of cells from the range of substrate surfaces can be seen in Figure 2. Cells were most easily detached from Ti6Al4V with only its native oxide present by both normal and shear forces. In general it was easiest to remove cells using normal forces rather than shear forces by this method.

Initially, at low stresses a greater proportion of chondrocytes were removed from PEO(A) compared to the other two types of PEO. However at higher stresses greater proportions of chondrocytes were removed from the other PEO surfaces. After the initial detachment of around 50 % of the cells adhered to PEO(A) by relatively low forces, few further cells are removed until much higher forces were applied. The reason for the ease of removal of a high proportion of cells from this surface is likely to be linked to the higher proportion of chondrocytes which maintained a rounded morphology for longer when seeded onto the PEO(A), Figure 1d. It is likely to be these cells which are removed even by low forces and the cells which have spread further on the surface, which adhere more strongly. Why spread cells adhered more strongly to this surface compared to the other surfaces may be due to the chemistry of the surface, but more likely the topology as similar phase are present in PEO(M). The PEO(A) surface is the roughest of the coatings and its topology is less ordered that that of PEO(M) surface.

When normal forces were applied to remove osteoblasts, it was shown that the strength of adhesion of cells to uncoated Ti6Al4V and PEO(P) was weaker than the strength of adhesion to PEO(M) and PEO(A). This is likely to be due to the topography of the surfaces. PEO(A) and PEO(M) were much rougher than the other surfaces and roughness is known to often increase cell adhesion. There may also be a contribution from the chemistry of the surfaces. In contrast to when normal forces were used, there was no statistically significant difference in shear forces required to remove osteoblasts from the different surfaces.

This study aims to show the potential for the use of centrifugation as a method of determining adhesive strengths of cell populations and consequently as a tool for use in assessing the biomedical potential of newly designed coatings.

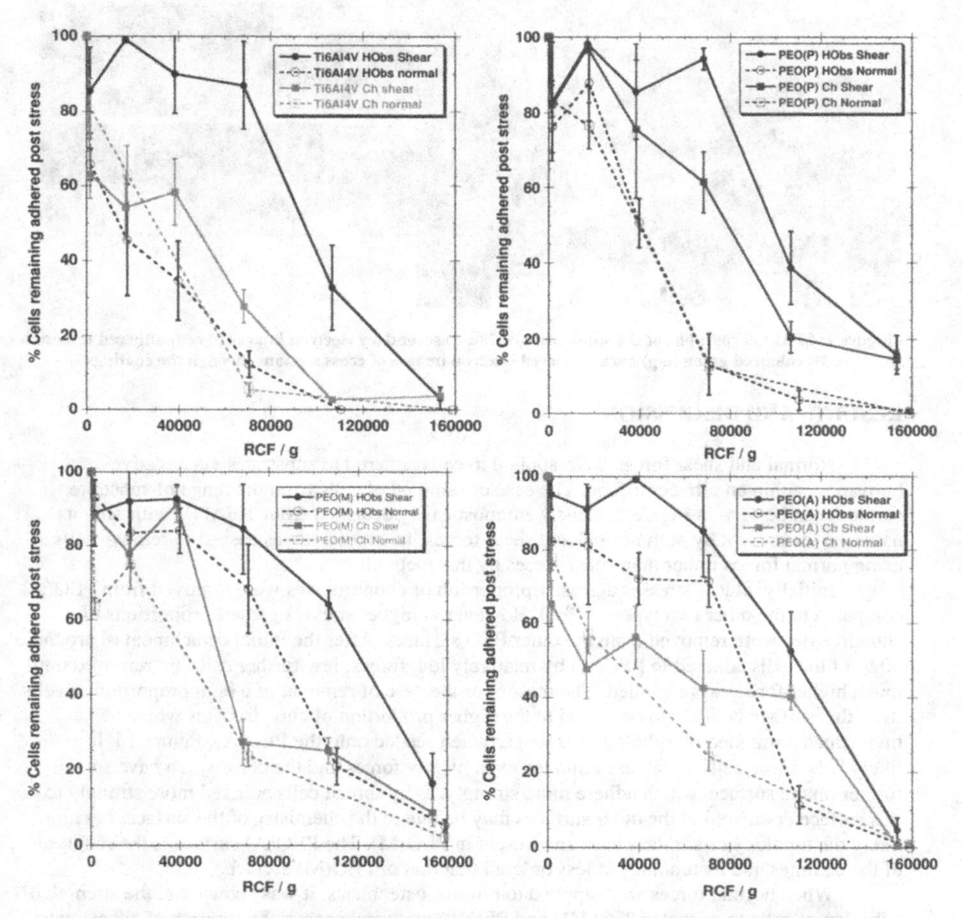

Figure 2: Graphs showing the ease of cell removal from the range of substrates. Both chondrocytes and osteoblasts were easier to remove using normal forces than shear forces.

CONCLUSIONS

- When stresses are applied via centrifugation it is easier to remove osteoblasts and chondrocytes using normal forces than shear forces.
- Chondrocytes adhere more strongly to all the tested PEO surfaces compared to Ti6Al4V with only a native oxide coating.
- Chondrocytes display different adhesion behaviour to PEO surfaces formed in different electrolytes and it may be possible to further tailor the chemistry, pore architecture and surface topography to optimise cell adhesion to them.
- The shear strength of the osteoblast-substrate interface remained constant for the range of surfaces tested, but the normal strength increased with roughness.

ACKNOWLEDGMENTS

Many thanks to the EPSRC for funding.

REFERENCES

[1] B. D. Ratner, Hoffman, A.S., Schoen, F.J., Lemons, J.E., "Biomaterials Science: An Introduction to Materials in Medicine," 2nd Edition ed: *Elsevier Academic Press*, 2004.

[2] L. H. Li, H. W. Kim, S. H. Lee, Y. M. Kong, and H. E. Kim, "Biocompatibility of titanium implants modified by microarc oxidation and hydroxyapatite coating," *Journal of Biomedical Materials Research*, vol. A 73A, pp. 48-54, 2005.

[3] F. Liu, F. Wang, T. Shimizu, K. Igarashi, and L. Zhao, "Formation of hydroxyapatite on Ti-6Al-4V alloy by microarc oxidation and hydrothermal treatment," *Surface & Coatings Technology*, vol. 199, pp. 220-224, 2005.

[4] S. H. Lee, H. W. Kim, E. J. Lee, L. F. Li, and H. E. Kim, "Hydroxyapatite-TiO2 hybrid coating on Ti implants," *Journal of Biomaterials Applications*, vol. 20, pp. 195-208, 2006.

[5] L. H. Li, Y. M. Kong, H. W. Kim, Y. W. Kim, H. E. Kim, S. J. Heo, and J. Y. Koak, "Improved biological performance of Ti implants due to surface modification by micro-arc oxidation," *Biomaterials*, vol. 25, pp. 2867-2875, 2004.

[6] D. Wei, Y. Zhou, D. Jia, and Y. Wang, "Characteristic and in vitro bioactivity of a microarc-oxidized TiO_2-based coating after chemical treatment," *Acta Biomaterialia*, vol. 3, pp. 817-827, 2007.

[7] J. P. Schreckenbach, G. Marx, F. Schlottig, M. Textor, and N. D. Spencer, "Characterization of anodic spark-converted titanium surfaces for biomedical applications," *Journal of Materials Science: Materials in Medicine*, vol. 10, pp. 453-457, 1999.

[8] J.-Z. Chen, Y.-L. Shi, L. Wang, F.-Y. Yan, and F.-Q. Zhang, "Preparation and properties of hydroxyapatite-containing titania coating by micro-arc oxidation," *Materials Letters*, vol. 60, pp. 2538-2543, 2006.

[9] J.-H. Ni, Y.-L. Shi, F.-Y. Yan, J.-Z. Chen, and L. Wang, "Preparation of hydroxyapatite-containing titania coating on titanium substrate by micro-arc oxidation," *Materials Research Bulletin*, vol. 43, pp. 45-53, 2008.

[10] J. A. Curran and T. W. Clyne, "Porosity in Plasma Electrolytic Oxide Coatings," *Acta Materialia*, vol. 54, pp. 1985-1993, 2006.
[11] H. J. Griffiths, J. G. Harvey, J. Dean, J. A. Curran, A. E. Markaki, and T. W. Clyne, "Characterisation of Cell Adhesion to Substrate Materials and the Resistance to Enzymatic and Mechanical Cell-Removal," *MRS Proceedings*, vol. 1097E, pp. GG03-05, 2008.
[12] "alamarBlue™ Technical Datasheet," AbD Serotec Ltd, 2002.

Mater. Res. Soc. Symp. Proc. Vol. 1191 © 2009 Materials Research Society 1191-OO05-12

Artificial Cilia: Mimicking Nature Through Magnetic Actuation

S. N. Khaderi[1], M. G. H. M. Baltussen[2], P. D. Anderson[2], D. Ioan[3], J. M. J. den Toonder[2] and P. R. Onck[1]
[1]Zernike Institute for Advanced Materials, University of Groningen, Nijenborgh 4, 9747AG Groningen, The Netherlands
[2]Eindhoven University of Technology, Den Dolech 2, 5612 AZ Eindhoven, The Netherlands
[3]Politehnica University of Bucharest, Spl. Independentei 313, 77206 Bucharest, Romania

ABSTRACT

Manipulation of bio-fluids in microchannels faces many challenges in the development of lab-on-a-chip devices. We propose magnetically actuated artificial cilia which can propel fluids in microchannels. These cilia are magnetic films which can be actuated by an external magnetic field, leading to an asymmetric motion like that of natural cilia. The coupling between different physical mechanisms (magnetostatics, solid mechanics and fluid dynamics) is numerically established. In this work we quantify the flow through a microfluidic channel as a function of its geometry for a characteristic set of dimensionless parameters.

INTRODUCTION

Manipulation of bio-fluids in microchannels finds very interesting applications in medical diagnostics. For example, in lab-on-a-chip devices a bio-fluid which is to be analyzed has to pass through several stages, such as mixing (with a catalyst), incubation (during which reactions take place over precise time periods) and detection of biomolecules. The search for new ways of propelling the bio-fluid through these stages is a very active field of research [1]. One of the characteristics of fluid flow at the micron scale is that the viscous forces are dominant over the inertial forces so that the latter can be considered negligible. A consequence of this is that the movement of the actuator should be asymmetric [2] in order for the fluid to be propelled. Such actuators can be found in nature (i.e. hair-like structures known as cilia) which enable micro organisms, such as paramecia, to swim. Recently we have demonstrated that magnetic artificial cilia, actuated by an external magnetic field, are able to propel fluids in microchannels [3].

In this work [3], the natural cilia are mimicked by polymer films with embedded nanoparticles which can be actuated by a tuned external magnetic field to exhibit an asymmetric motion. Two configurations were proposed: a permanently magnetic film (PM) subjected to a sudden magnetic field and a super-paramagnetic film (SPM) under the influence of a rotating external magnetic field. It was shown that the flow through a microfluidic channel is proportional to the area swept by the film, and we explored the dependence of the swept area on the ciliated-system parameters (film length and thickness, mechanical and magnetic film properties and the applied field) for one specific channel geometry. In this paper, however, we fix the parameters of the cilia system and explore the effect of the channel geometry (cilia spacing and height) on the induced fluid flow.

METHOD

We study a periodic arrangement of cilia in an infinitely long channel. The computational unit-cell that we analyze is shown in Figure 1. The spacing between the cilia (i.e. the horizontal size of the unit-cell) is denoted by W and the channel height by H. For a given unit cell, the dimensionless parameters which govern the system are: the inertia number $I_n=12\rho L^4/Eh^2t_{ref}^2$, the fluid number $F_n=12\eta L^3/E/h^3t_{ref}$ and the magnetic number $M_n=12N_zL^2/Eh^2$, where L, h, ρ and E are the length, thickness, density and elastic modulus of the cilium, respectively, η is the viscosity of the fluid, N_z is the magnetic body couple and t_{ref} is the time during which the magnetic field is applied. Here we take the value of these dimensionless numbers the same as in [3], except for M_n in the case of PM film, where its value is twice that in [3]. When these numbers are kept constant the area swept by the film, hence the performance of the cilia, will remain the same, for a given channel geometry. The objective of the current paper is to describe the performance of the microfluidic device for these I_n, F_n and M_n as a function of the channel geometry (W/L, H/L).

To perform the parametric study of flow across such a microfluidic channel as a function of its geometry, a fully coupled solid-fluid interaction model is developed. The inertia of the fluid is neglected as the microfluidic channel is likely to operate at low Reynolds numbers. Hence, the fluid motion is governed by the Stokes equation. The magnetic film is modeled as an assemblage of Euler-Bernoulli beam elements taking into account the inertia and geometric nonlinearity of the film. The solid and the fluid models are coupled with a constraint that the velocity of the film and the fluid are identical at discrete points on the film. A detailed description of the fluid structure interaction model can be found in [4].

RESULTS

As can be seen from Figure 1, the film is kept at the bottom of the channel, but slightly away (1/10 of its length) from the substrate. This is done to avoid any stiction problem in the case of a SPM film and to avoid contact with the substrate in the case of SPM and PM film. As a consequence we can observe a small 'leakage' flow between the film and the substrate. From the velocity contours, it can be seen that this leakage flow is very small when compared to the main flow. No-slip boundary conditions are applied at the top and bottom of the channel and on the left and right ends of the channels, we prescribe periodic boundary conditions. This makes the pressure at the left and right ends of the channel to be the same, so that the fluid flow that we observe is solely due to the pressure gradient generated around the individual cilia. Both the SPM and PM film are analysed. In the analysis I_n, F_n and M_n are kept the same. Only the channel geometry is changed.

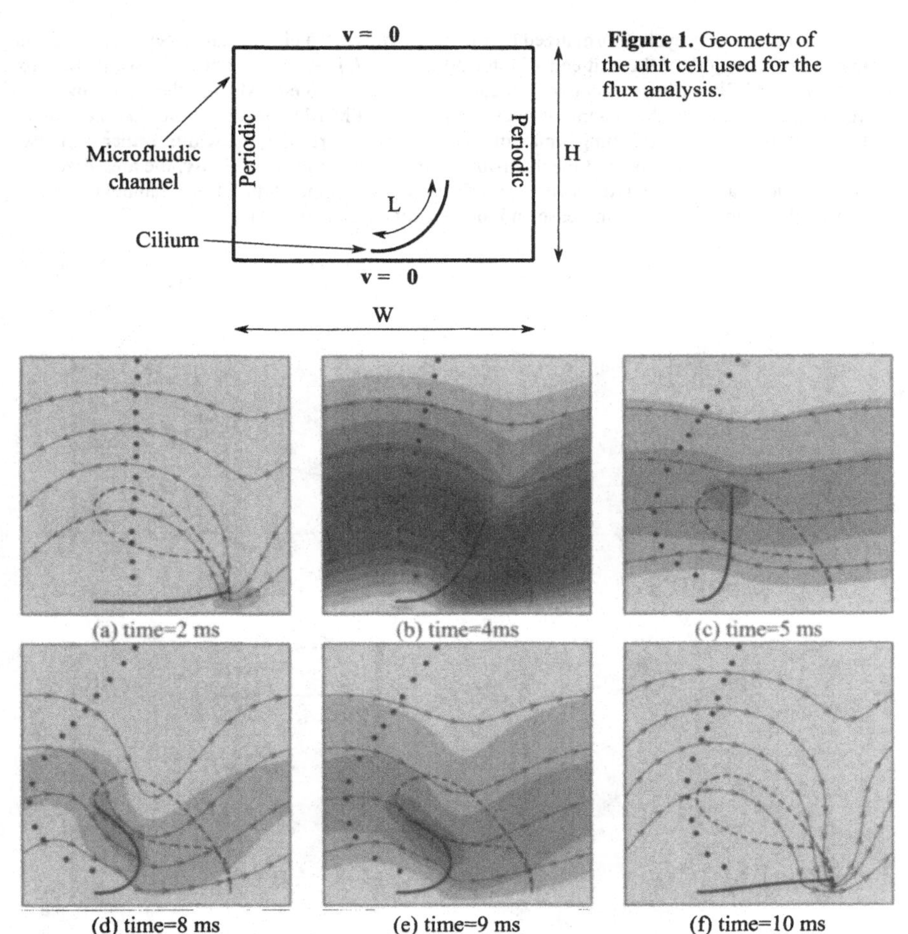

Figure 1. Geometry of the unit cell used for the flux analysis.

(d) time=8 ms (e) time=9 ms (f) time=10 ms

Figure 2. Velocity field and streamlines at different times for a SPM film. The solid black dots at different instances represent the position of fluid particles. The dashed lines represent the trajectory. The contours represent the absolute velocity normalized by L/t_{ref}, which take a maximum value of 4.5 and a minimum of zero.

As an example, the velocity field at different instances of time for an SPM film for $H/L=2$ and $W/L=2$ is shown in figure 2. The pumping action can be nicely seen from the motion of fluid particles due to the motion of the cilia. The fluid particles are represented by solid black dots. If we look at the first and the last instances it can be seen that the fluid particles have suffered a net displacement to the left during the cycle.

The area flow per cycle (normalized by the maximum area a cilium can sweep, i.e. $\pi L^2/2$) as a function of the width of the unit cell W/L for different H/L is shown in figures 3(a) and 3(c), for the PM film and SPM film, respectively. It can be seen that for the SPM film, the area flow increases as the width of the channel is decreased. For the PM film, however, the flux increases as the width is decreased, reaching a maximum at $W/L=4$. Decreasing the width further decreases the area flow. We see two distinct mechanisms for the decrease in flux. Firstly, the area flux decreases due to a decrease in the swept area. Secondly, when the width of the channel is increased the fluid drag forces increase and thus the area flux gets reduced.

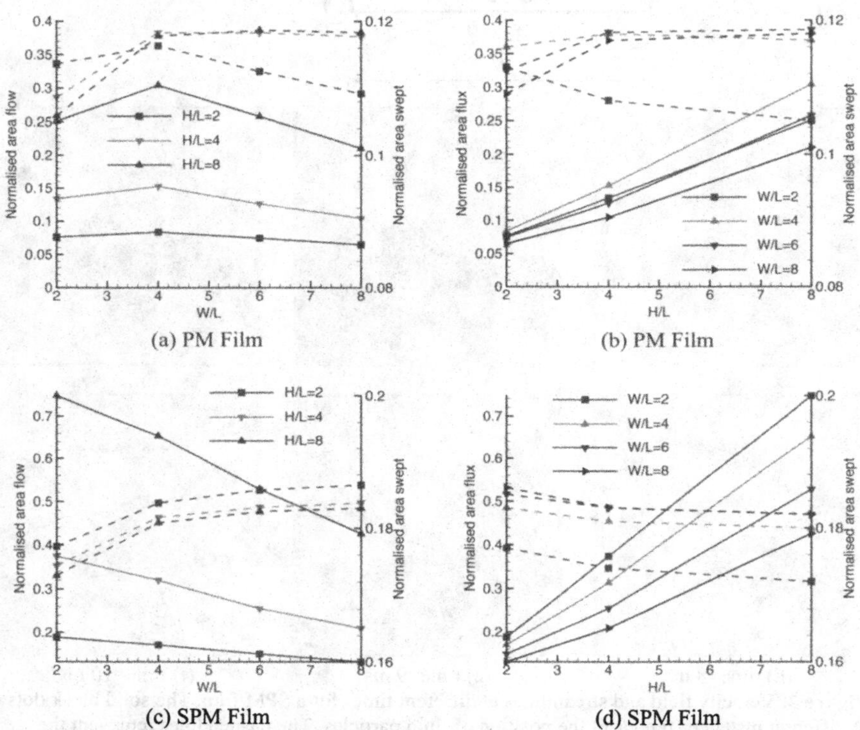

Figure 3. Area flow as a function of geometry. Solid lines represent the area flow out of the channel (left axis) and the broken lines represent the area swept by the cilium (right axis).

As the SPM film is subjected to a rotating magnetic field, its velocity is nicely controlled by the rate of the applied field. As a result, the film sweeps approximately the same area. This can be observed when we see the trajectory of the free end of the film as the width is increased, figures 3(c) and 4(b). In fact it, can be seen from figure 3(c) that the swept area shows a very small increase. Hence, the reduction in the area flow (as we increase the width) is mainly due to increased fluid viscous forces, due to the larger cilia spacing W in the channel.

In the case of a PM film we observe this mechanism only when the width is large ($W>4L$). In the case of a PM film we apply a constant magnetic field for a certain duration making the film to buckle after which the film elastically restores to the initial position. Hence, the velocity of the film is not controlled by the applied magnetic field as in the case of a SPM film, but is controlled by the viscous forces (also by the elastic forces, but the elastic properties of the film are kept constant), which decrease when the width of the unit cell is reduced. Hence it should be expected that, as the velocity of the film is high for a low cell width (due to the reduced viscous forces), the film should sweep larger areas. However, the opposite is the case. As the width of the unit-cell is decreased, the fluid drag forces on the film are reduced and the film shows a different motion (figure 4(a)) which leads to a distinct reduction in the swept area (figure 3(a)). Hence, the area flow reduces when the width of the channel is decreased below $4L$.

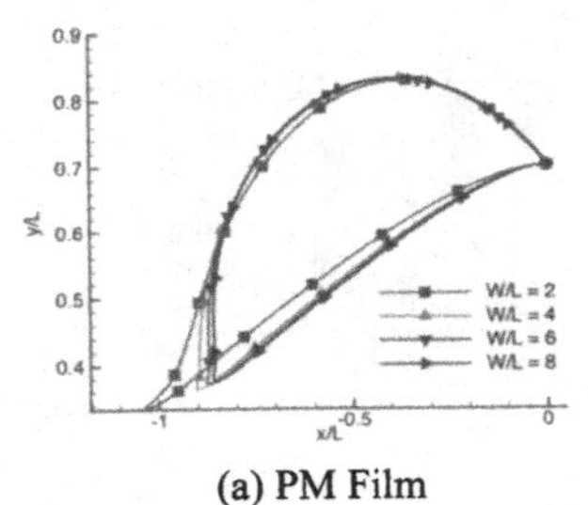

(a) PM Film

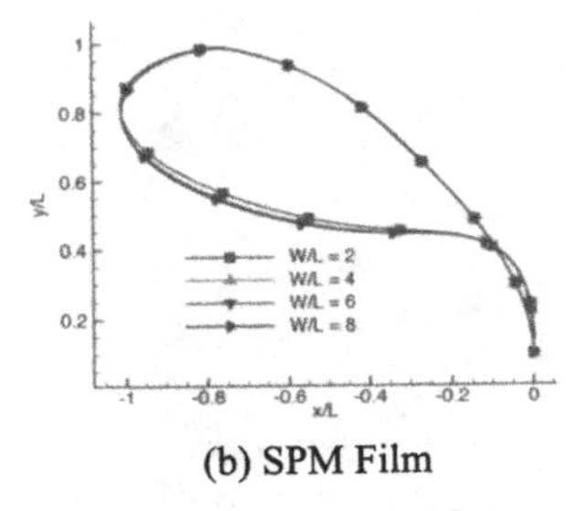

(b) SPM Film

Figure 4: Trajectory of the free end for *H/L=4*.

The same data of figures 3(a) and 3(c) can be plotted as a function of *H/L* to elucidate another interesting behavior. For a given *W/L*, we can see that the flux increases linearly with *H/L* for both PM and SPM film. The reason can be deduced from the velocity contours and the respective velocity profiles at the periodic unit-cell boundary. Figure 5 shows this for a SPM film. The maximum velocity occurs at the free end of the film and the velocity is zero at the bottom and the top of the channel. As *H/L* is increased the velocity profile becomes nearly linear from $y=L$ to $y=H$. In addition, as *H/L* is increased the maximum normal velocity (u) at $y = L$ slightly increases. Hence, the overall area flow increases as *H/L* is increased, see figures 3(b) and 3(d).

ACKNOWLEDGMENTS

This work is a part of the 6th Framework European project 'Artic', under contract STRP 033274.

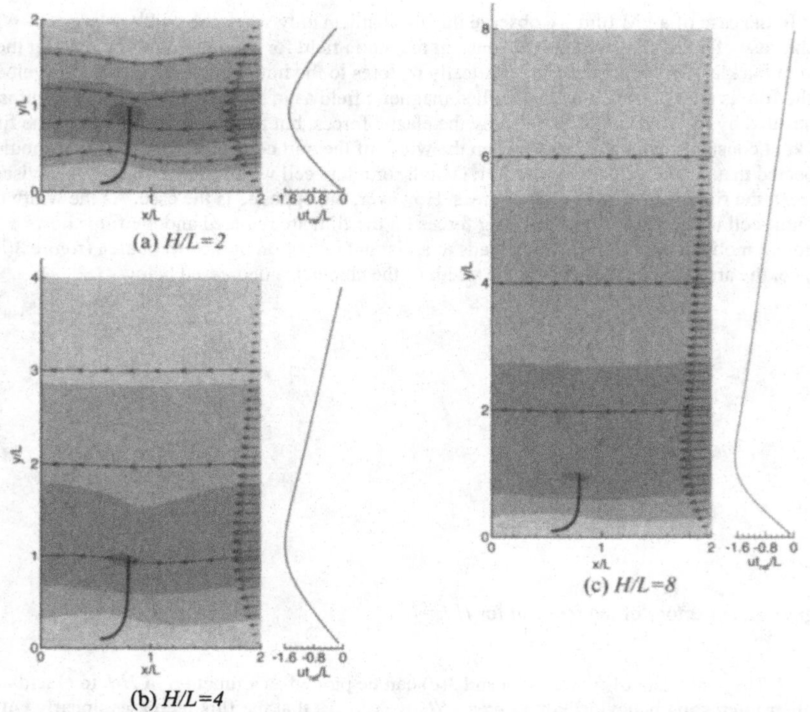

Figure 5: SPM film: Velocity profiles at 5 ms for $W/L=2$. Variation of the horizontal component of velocity at the right boundary along the height of the channel at $x/L=W/L$ is plotted next to the respective velocity profile. The maximum velocity occurs at the free end of the film and zero at top substrate and top of the channel. As H/L is increased the velocity profile becomes nearly linear at the exit of the channel. Moreover, as H/L is increased the normal velocity (u) at the exit of the channel increases. Hence the area flow increases as H/L is increased, figure 3(d). The contours represent the absolute velocity normalized by to L/t_{ref}, which take maximum value of 4.5 and a minimum of zero.

REFERENCES

1. D. J. Laser, and J. G. Santiago, *Journal of Micromechanics and Microengineering* **14**, R35 (2004).

2. E. M. Purcell, *American Journal of Physics* **45**, 3 (1977).

3. S. N. Khaderi, M. G. H. M. Baltussen, P. D. Anderson, D. Ioan, J. M. J. den Toonder and P. R. Onck, *Phys. Rev. E* **79**, 046304 (2009).

4. R. van Loon, P. D. Anderson, F. N. Van de Vose and S. J. Sherwin, *Computers and Structures* **85**, 833 (2007).

Advances in Device Materials

Mater. Res. Soc. Symp. Proc. Vol. 1191 © 2009 Materials Research Society 1191-OO06-01

Lab-on-Glass System for DNA Analysis Using Thin and Thick Film Technologies

D. Caputo[1], M. Ceccarelli[1], G. de Cesare[1], A. Nascetti[2], R. Scipinotti[1]
[1]Dept. of Electronic Engineering, "Sapienza" University of Rome, Rome (Italy)
[2]Dept. of Aerospace and Astronautics Eng., "Sapienza" University of Rome, Rome (Italy)

ABSTRACT

In this paper, we present a compact lab-on-chip system suited for label-free DNA analysis. The system can be fabricated on a conventional microscope glass slide using thin-film and thick-film technologies. It integrates a heating chamber, an electrowetting-based droplet handling system and a hydrogenated amorphous silicon (a-Si:H) photosensor array for DNA detection. At this stage of research we have designed and tested the individual functional units. The heating chamber incorporates a thin metal film heater optimized for uniform temperature distribution on a $1cm^2$ area. A forward-biased a-Si:H p-i-n junction is used for temperature monitoring, achieving a linear temperature dependence with -3.3 mV/K sensitivity. The droplet-handling unit, relying on the electrowetting method, is designed to move the sample from the heating chamber to the sensor array. The unit includes a set of metal pads beneath a layer of PDMS that provides both the electric insulation of the electrodes and the hydrophobic surface needed by the electrowetting technique. The UV sensor array allows measuring the DNA absorbance variation at 254nm related to the hybridization between probe-molecules contained in the sample and reference target molecules immobilized on the sensor surface. A preliminary test to detect the hybridization between a 25-mer single-stranded oligonucleotides and denaturated pBR 322 4162-mer single-stranded oligonucleotides has been carried out successfully.

INTRODUCTION

A variety of recent technological breakthroughs in molecular biology and microfabrication technology have made possible the development of lab-on-chip (LOC) systems. The high integration level of the LOCs allows to accomplish complex chemical or bio-chemical functions of large analytical devices on a single sensor-like system with a fast response time, low sample consumption and on-site operation [1][2]. The functional modules included in LOC systems are those capable of sample injection, reaction, separation and detection. In the framework of LOC systems the most promising are the DNA Chips (or DNA microarray) [3][4], which are already commercially available. Usually, DNA detection is performed using an off-chip detection systems able to measure the fluorescence emitted by fluorescent dyes attached to the target DNA molecules (labeled DNA) [5]. In alternative, label-free methods relying on the measurement of changes in electrical charge [6], mass [7] or UV-absorbance [8, 9] have been proposed.

In this paper we propose a LOC system able to perform pre-treatment, handling and detection of DNA molecules on a standard microscope slide fabricating by means of thin and thick film technologies. In particular, an on-chip detection is achieved by using hydrogenated amorphous silicon (a-Si:H) photosensors directly deposited on the glass. We refer to the whole system as lab-on-glass (LOG).

LAB-ON-GLASS SYSTEM CONCEPT

The presented system is composed by three main parts. Each part has a specific function in the sample-treatment chain as sketched in Figure 1. The pre-treatment unit provides the thermal control by means of a thin film resistor acting as heater and an amorphous silicon diode acting as temperature sensor. The sample-handling unit has the task to move the samples from the pre-treatment area to the detection unit, where the biomolecules are detected by using an a-Si:H sensor array.

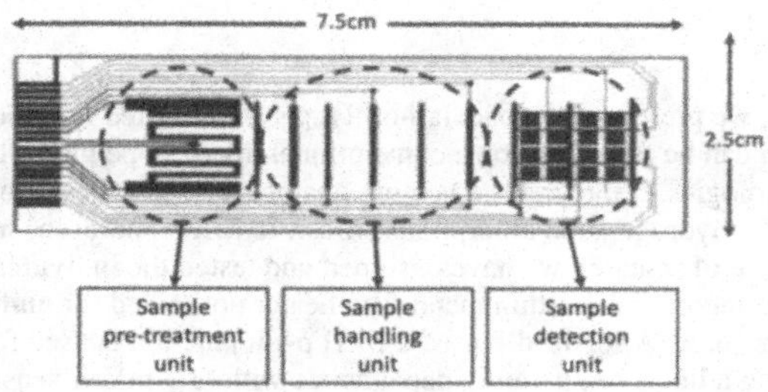

Figure 1. Structure of the Lab-on-Glass.

In this work, the three parts of the presented LOG have been studied as individual units taking into account the final project to integrate all of them in a single device. According to this specification, each functional block has been fabricated considering the requirements and compatibility of all the technological steps.

SAMPLE PRE-TREATMENT UNIT

In many applications pre-treatment involves a thermal cycling of the sample e.g. for DNA amplification by Polymerase Chain Reaction (PCR). In our system this function is implemented by a thin film heater, a temperature sensor and a PDMS chamber. In order to get a spatial-uniform temperature distribution over the whole active area the heater geometry has been designed using multiphysics finite element simulations (COMSOL Multiphysics), which couples the electrostatic problem and the heat transfer problem. We optimized the width and the spacing of the segments forming a serpentine-shaped resistor achieving a temperature distribution with uniformity better than 2% over the entire 1 cm^2 area. (solid line in Figure 2a). The heater has been fabricated with a 2000 Å thick Ti/W film resistor deposited by magnetron sputtering on the glass substrate. The active area of the heater is 1 cm^2. The measured resistivity is $3.8 \cdot 10^{-4}$ Ω/cm. In Figure 2b we report, as symbols, the measurements of the temperature distribution performed using a thermo-camera (AVIO NEOThermo TVS620P). The experiment has been performed applying a voltage of 30 V causing a maximum temperature of 90°C. The measured uniformity has been found to be better than 3% confirming the modeled results.

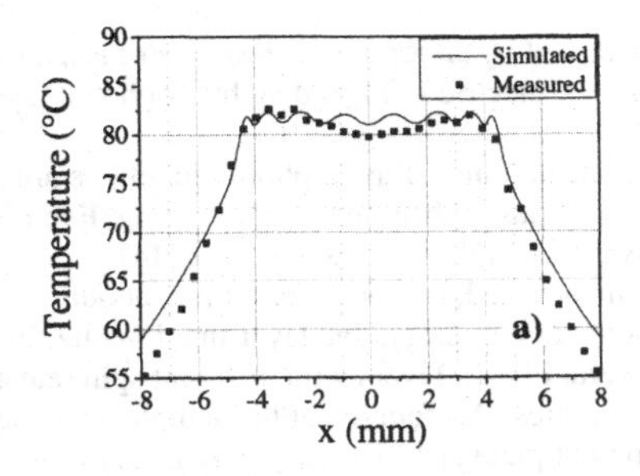

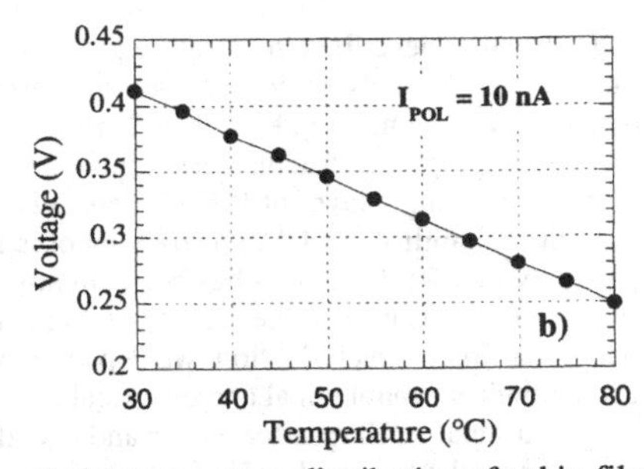

Figure 2. a) Simulated (line) and measured (symbols) temperature distribution of a thin-film resistor. b) Measured voltage across the a-Si:H diode, biased with 10nA constant current, as a function of temperature.

In order to integrate temperature sensing in the structure, the suitability of an amorphous silicon diode as temperature sensor has been studied. For its fabrication, a 2000 Å thick film of Ti/W has been at first sputtered on the glass substrate as bottom contact of the device. In the complete LOG structure this electrode will to be deposited simultaneously with the thin film heater. The sensor is a n-type a-Si:H/i-type a-Si:H/p-type a-SiC:H (amorphous silicon carbide) stacked structure deposited by Plasma Enhanced Chemical Vapor Deposition (PECVD). The top contact has been ensured by a 2000 Å thick Ti/W metal layer. A square-shaped diode with area of $1mm^2$ has been characterized performing current-voltage (I-V) measurements at different temperatures ranging from 20°C up to 80°C with step of 5°C. We found that, in forward bias conditions, at constant bias current, the voltage across the diode is linearly dependent on the temperature. In particular, at 10 nA constant bias current, a sensitivity of -3.3 mV/K, as shown in Figure 2b, has been achieved. We have observed that this value does not significantly change with thicknesses of the a-Si:H layers. Therefore in order to reduce and to keep the technological steps as simple as possible, we deposited the a-Si:H diode with the same deposition parameters utilized for the UV photosensor described in the detection unit.

The confinement of the sample above the heater has been obtained fabricating a PolyDiMetylSiloxane (PDMS) chamber. The chamber has been bonded to the microscope slide by exposing the surfaces of both substrate and chamber to an oxygen plasma (40W, 3 minutes, 200mT, 100sccm). The water-tightness of the fabricated PDMS chamber has been verified introducing 100 μl of water inside the chamber and placing it on a hot plate heated at 100°C: after cooling down at room temperature the entire water sample has been removed from the camera confirming the absence of any leakage.

SAMPLE HANDLING UNIT

The handling unit is used to move DNA sample droplets from the heating chamber to the detection unit. The movement is achieved implementing the electrowetting-on-dielectric (EWOD) method. The EWOD relies on the possibility to change the contact angle of a liquid droplet in contact with a hydrophobic layer by means of an electric fields generated by an insulated control electrode, according to the Young-Lippmann equation [10][11]. The typical

EWOD structure described in literature [12] is constituted by an array of metal electrodes, an insulation layer (usually silicon nitride) that sustains the electric field, and an hydrophobic layer (typically PTFE, commonly known as Teflon).

In the proposed structure (see Figure 3a) the insulation and hydrophobic functions are performed by a single layer of PDMS, deposited on the glass substrate by spin coating. Even though, the suitability of PDMS as hydrophobic layer [13] and as insulation layer [14] respectively in EWOD devices has been already demonstrated, this is the first time, at our knowledge that it is used at the same time as insulation and hydrophobic layer in a EWOD device. As a first characterization, we deposited several PDMS layers with different spin rate and measured their morphological and electrical characteristics. We found that the sample deposited by spin coating at 6000 rpm for 30 sec and baked on hot plate at 165°C for 20 minutes in vacuum shows the best trade-off between smoothness of surface (which decreases with rpm) and thickness of the layer (which decreases with rpm). A better smoothness ensures a easier movement while a thinner sample allows the use a lower voltage to obtain the droplet movement. For this sample, a breakdown voltage greater than 20 MV/m and a contact angle around 117° have been achieved.

Starting from these results, we fabricated a EWOD structure consisting of an array of 1 mm^2 area squared Ti/W electrodes deposited by sputtering at 100°C on the glass substrate covered by a layer of PDMS deposited with the recipe reported above. The resulting PDMS thickness is 1 μm. The area of the electrode is enough to contain the 2μl water droplet typically used in our biomolecular detection experiments[15]. The gap between electrodes is 75 μm wide. This distance has been chosen as an average value between the data reported in literature [14].

We performed the experiment using a 2 μl of water and applying voltages, between adjacent electrodes, up to 200V. In Figure 3b it is visible the deformation of the drop caused by the applied voltage. Even though movement is achieved starting from 75Volts, only voltages above 100Volts demonstrated a reproducible movement of the droplet. No damage of the PDMS layer has been observed up to 200 Volts.

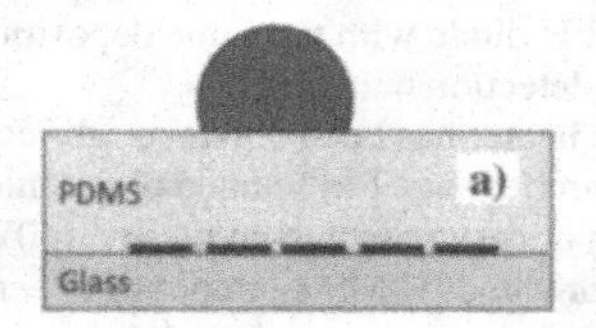

Figure 3. a) Cross section of the EWOD device. b) Frame showing the deformation of the droplet during the movement.

SAMPLE DETECTION UNIT

On-chip DNA detection has been performed by using a-Si:H photosensors whose structure has been optimized to maximize the responsivity in the UV range. In particular, the device is able to detect the different absorbance of single and double stranded DNA molecules. The a-Si:H layers of the device, depicted in Figure 4a, have been deposited with the parameters reported in Table I. A Cr/Al/Cr stacked layer metal layer acts as bottom electrode. However, the choice of chromium and aluminum as metal electrode is not restrictive. According to the requirements of the pre-treatment unit and of the droplet handling unit, Ti/W can be used as the metal for the electrodes without impairing the performances of the sensor.

Table I. PECVD parameters used in the deposition of the a-Si:H and a-SiC:H layers of the sensor structure. The gases are: SiH_4 pure silane, PH_3 silane diluted (5%), B_2H_6 helium diluted (5%), CH_4 pure methane; P_D is the process pressure; P_{RF} is the power density of the plasma discharge; T_D is the substrate temperature; t_D is the deposition time.

Layer type/material	SiH_4 (sccm)	PH_3 (sccm)	B_2H_6 (sccm)	CH_4 (sccm)	P_D (Torr)	P_{RF} (mW/cm^2)	T_D (°C)	t_D (sec)	Thickness (nm)
n / a-Si:H	40	10			0.3	25	200	180	30
i / a-Si:H	40				0.3	25	180	900	150
p / a-SiC:H	40		3	60	0.3	25	160	180	5

The top electrode is an Al/Cr metal grid, whose spacing was optimized for a 300 μm charge collection length according to the conductivity of the underlying p-layer [16]. The area of the fabricated sensors is 2x2 mm and the metal grid has a pitch of 200 μm with 50 μm wide fingers. The measured sensor responsivity is around 45mA/W at 254 nm.

An hybridization experiment has been performed to test the ability of the sensor to distinguish between single- and double-stranded DNA molecules. The sample was prepared immobilizing a 25-mer single-stranded DNA (probe) on a functionalized quartz substrate. Half of the quartz substrate has been then exposed to linearized and denaturized pBR 322 4162-mer single-stranded oligonucleotides (target).

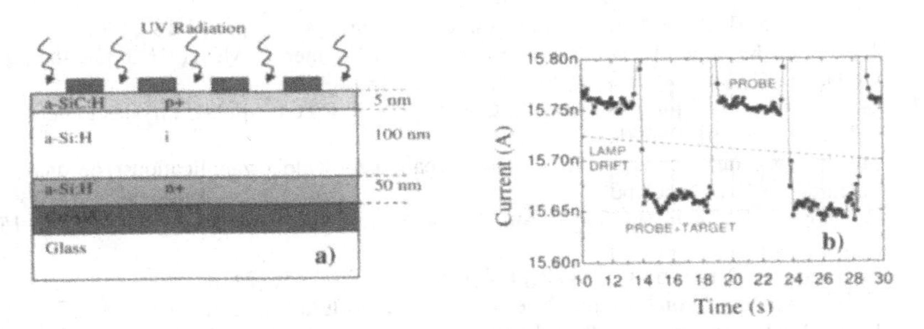

Figure 4. a) Structure of the UV sensor. b) Sensor photocurrent as function of time for single strand (PROBE) and hybridized (PROBE+TARGET) DNA.

The measurements have been performed by alternating the two halves of the quartz substrate (namely the one with only the probe DNA and the one with the hybridized probe and target) in the optical path of a 254nm UV radiation impinging on the a-Si:H sensor. Results of Figure 4b, showing the current of the UV sensor as function of the time, demonstrated the successful detection of the hybridization process. The long term drift observed in the figure has to be ascribed to the instability of the radiation source.

For the final LOG system we propose the use of SiO_2 as passivation layer of the array instead of or in addition to the Si_3N_4 layer because the surface functionalization for DNA immobilization is not effective on silicon nitride [17].

CONCLUSIONS

We have presented a DNA analysis system for Lab-on-Glass applications. The system is composed by three parts that have been studied, fabricated and characterized separately taking into account the requirements for a future integration on the same glass substrate. A thin film heater with an integrated p-i-n a-Si:H diode constitute the pre-treatment unit. An array of sputtered Ti/W electrodes allows to move droplets of the solution containing the DNA from the heating chamber to the detection section by using EWOD methods. An array of a-Si:H UV sensor is used to detect the hybridization of DNA molecules. In particular, the possibility to distinguish single layers of single- and double-stranded DNA molecules immobilized on a quartz surface has been proven.

REFERENCES

1. Kopf-Sill AR (2002), "Successes and challenges of lab-on-a-chip", Lab Chip 2:42N–47N
2. Manz, A.; Graber, N.; Widmer, H. M. Sens. Actuators 1990, B1, 244-248.
3. S.K. Moore, "Making Chips", IEEE Spectrum, pp. 54-60, (2001)
4. Eric T Lagally and Richard A Mathies, , J. Phys. D: Appl. Phys. 37 (2004) R245–R261
5. D. Meldrum, Genome Res., 21, pp- 20-24, (1999)
6. C. Berggren, B. Bjarnason, G. Johansson, Electroanalisys 13, pp. 173-180, (2001)
7. J. Fritz, M. K. Baller, H. P. Lang, H. Rothuizen, P. Vettiger, E. Meyer, H.-J. Guntherodt, Ch. Gerber, J. K. Gimzewski, Science 288, pp. 316-318 (2000)
8. G. de Cesare, D. Caputo, A. Nascetti, C. Guiducci, B. Riccò, Applied Physics Letters, (2006), vol. 88, pp. 083904
9. http://www.bmglabtech.com/db_assets/applications/downloads/applications/168-dna-quantitation-absorbance.pdf
10. B. Berge, C. R. Acad. Sci. Ser. II: Mec., Phys., Chim., Sci. Terre Univers, 1993, 317, 157–163.
11. M. Vallet, B. Berge and L. Vovelle, Polymer, 1996, 37, 2456–2470.
12. Roland Baviere, Jerome Boutet, Yves Fouillet, Microfluid Nanofluid (2008) 4:287–294.
13. Ding, Hui-Jiang; Liu, Kan; Zhao, Li-Bo; Zeng, Qian; Guo, Zhi-Xiao; Guo, Feng; Zhao, Xing-Zhong, Bioinformatics and Biomedical Engineering, 2007. ICBBE 2007. The 1st International Conference on Volume , Issue , 6-8 July 2007 Page(s):1317-1320.
14. Mohamed Abdelgawad and Aaron R. Wheeler, Adv. Mater, 1007, 19, 133-137.
15. D. Caputo, G. de Cesare, A. Nascetti, R. Negri, R. Scipinotti, IEEE Sensors Journal, vol. 7 (9), pp. 1274-1280 (2007)
16. D. Caputo G. de Cesare A. Nascetti M. Tucci, IEEE Trans. of Electron Device, vol.55(1), NUM:, p. 452-456 (2008).
17. F. Fixe, A. Faber, D. Goncalves, D. M. F. Prazeres, R. Cabeca, V. Chu, G. N. Ferreira and Conde J P 2002, Mater. Res. Soc. Symp. Proc. 723 O2.3.1

Mater. Res. Soc. Symp. Proc. Vol. 1191 © 2009 Materials Research Society 1191-OO06-06

Self-Assembled Ultra-Thin Silica Layers for On-Chip Chromatography

Sun Choi*,[1], Inkyu Park[2] and Albert P. Pisano[1]

[1]Berkeley Sensor and Actuator Center (BSAC), University of California at Berkeley, Berkeley, California 94720, USA

[2]Department of Mechanical Engineering, Korea Advanced Institute of Science and Technology (KAIST), 335 Gwahangno, Yuseong-gu, Daejon 305-701, South Korea

ABSTRACT

We propose a novel method for self-assembled packing of silica microsphere in micro-channel which can be potentially used for on-chip chromatography. Solvent-evaporation based 2-D crystallization technique [1] can enable mono-dispersed micro-particles to be self-assembled by capillary attractive forces. We apply this technique to assemble dense packing of silica microsphere and form ultra thin layers (2~3 layers) within open microchannel. Open micro-channel has been constructed by conventional photolithography of SU8 photoresist. A small droplet (Volume: 0.1μL) of silica suspension (Diameter: 3μm, Solvent: DI Water, Concentration: 1.25wt%) has been placed in the defined inlet of micro channel. Capillary force within the open SU8 microchannel induces the flow of silica suspension in the channel. The packing of microsphere starts from the outlet side of the channel, where the thickness of solvent drastically decreases due to sudden increase of cross-sectional area of channel, and this packing propagates to the inlet side of the channel until solvent evaporates completely. As a result, a dense packing of silica microspheres are successfully assembled and a thin layer of silica microspheres are formed within open micro-channel.

INTRODUCTION

Chromatography has been one of the most widely used techniques for the analysis and separation of the mixtures of biochemical compounds in research laboratories and industrial factories. Numerous chromatography techniques such as High-Performance Liquid Chromatography (HPLC), Thin Layer Chromatography (TLC) use absorbents (ex: silica, alumina, cellulose) as stationary phase material [2]. Effective loading of absorbents in those techniques has been a huge challenge since it requires additional implementation of high-pressure pump system (for HPLC) or limits selective coating of absorbents on supporting plate (for TLC). In order for chromatography to be efficiently integrated with micro-fluidic *Lab-on-a-chip* devices, novel techniques for easy and simple packing of absorbents within micro channels should be developed. We have demonstrated an ultra-fast microfluidic approach to self-assemble silica micro-particles in three-dimension by taking advantage of simple photolithography and capillary action of micro particles-dispersed suspensions. Silica and silica gel microspheres have been successfully assembled within micro-open channel by using this approach.

THEORY

Medium-evaporation based two dimensional crystallization technique [1] was applied to assemble mono-dispersed micro-particles of suspension in a micro-open channel which was fabricated by UV-photolithography. A small droplet of micro-particle dispersed suspension was placed in the defined inlet of channel to induce the assembly of micro particles within channel.

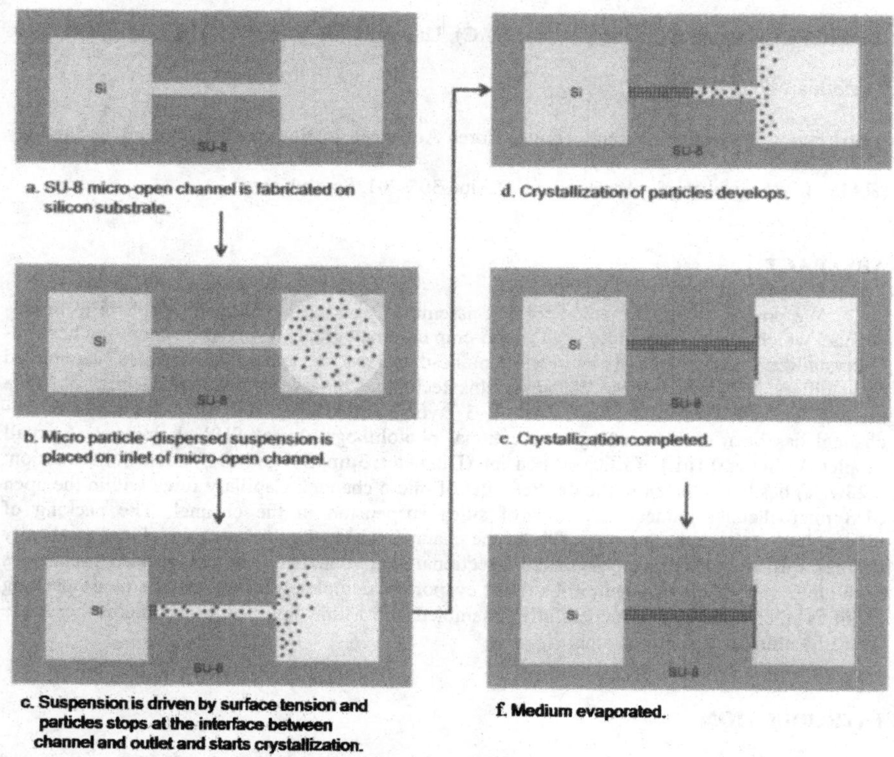

Si
SU-8
a. SU-8 micro-open channel is fabricated on silicon substrate.
b. Micro particle -dispersed suspension is placed on inlet of micro-open channel.
c. Suspension is driven by surface tension and particles stops at the interface between channel and outlet and starts crystallization.
d. Crystallization of particles develops.
e. Crystallization completed.
f. Medium evaporated.

Figure 1. Schematic of medium-evaporation based micro particle assembly in micro-open channel.

EXPERIMENT

Silica micro microspheres based suspension (Silicon dioxide based micro particles, diameter: 3 μm, 5 μm, water: 95 wt %, silica: 5 wt %, *Sigma-Aldrich*) has been diluted into various concentrations of suspension (2.5 wt %, 1.25 wt %, 0.625 wt %, 0.3125 wt % silica) by adding controlled volume of a deionized water. Dried silica gel spheres of which pore size is 60 Å, diameter is 3 μm and standard deviation is D10: 2.15μm, D50: 3.40μm, and D90: 6.42μm (SiliaSphere™, *Silicycle*) were mixed with controlled volume of deionized water to make suspension of 1.25 wt % silica gel and 98.75 wt % water. Micro-open channel has been fabricated by UV-Photolithography. Lightly doped p-type silicon wafer (*Silicon Quest International Inc.*) has been used for substrate and SU-8 2007 (*Microchem Corp.*) has been used for UV-sensitive negative photoresist. As can be seen from Figure S1, SU-8 resist was spincoated on top of silicon substrate and following photolithography and plasma oxygen treatment defined channel geometry and surface property of channel wall and substrate.

Two reservoirs and channel geometry has been designed as can be seen from Figure S1. For inlet / outlet reservoirs, 20mm × 20mm × 7μm (Width × Length × Height) rectangular patterns were defined to contain 0.2μL suspension volume without overflow. For channels, various widths (10μm, 13μm , 16μm , 20μm, 40μm, 60μm), various lengths (2mm, 3mm, 4mm, 5mm, 6mm, 7mm, 8mm) have been defined with the fixed height (7μm).

RESULTS & DISCUSSION

Silica microspheres were assembled tightly in both horizontal and vertical directions within the micro-open channel by using presented technique. Figure 2 shows that the particle assembly occurs not only in the direction of capillary flow but also vertical to the direction of flow. This observation supports that crystal growth of particles after initial nucleation is governed not only by capillary action of medium but also evaporation of medium between particles. Also, interface between channel and outlet reservoir has been sharply defined and no overflow of particles were being found, which is consistent with the claim that nucleation of particles begins where the thickness of water film is reduced to the diameters of particles. Commensurable effect of silica microspheres in micro-open channel was also demonstrated in more detail by varying the width of channel and the diameter of silica particles. Since crystal growth of particle assembly is progressed by the pathway of medium fluid, particle packing can be limited by fluid pathway. Horizontal and vertical scale of fluid is determined by original channel geometry thus commensurable effect can be found according to channel structure.

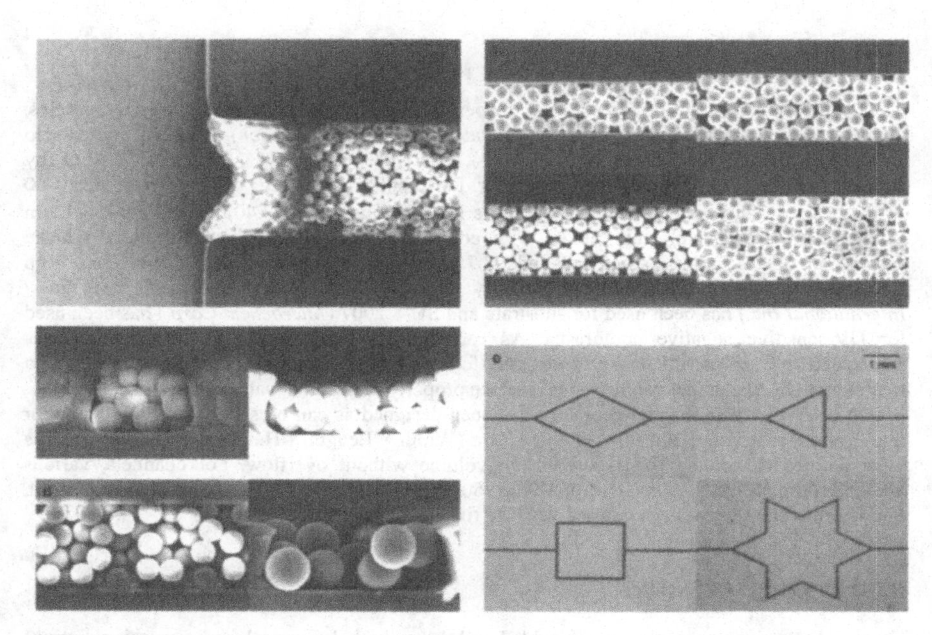

Figure 2. Images of assembly of 3 μm diameter- silica microspheres within SU-8 micro-open channel. **a-c,** (a) Interface between channel and outlet reservoir. Channel width is 40 μm, height is 7 μm and length is 3mm. Outlet reservoir is 2 mm × 2 mm × 7 μm (Height). Overflow of silica microspheres is not found and interface between silica and outlet reservoir is shapely defined. (b) Commensurable effect of silica microspheres in horizontal scale. The widths of channels are 10 μm (top left), 13 μm (top right), 16 μm (bottom left) and 20 μm (bottom right). The length (3 mm) and width (20 μm) of four micro channels are identical. The number of particles are 3 (top left), 4 (top right), 5 (bottom left), 6 (bottom right). (c) Commensurable effect of silica microspheres in vertical scale. The widths of channels are 10 μm (left) and 20 μm (right). The number of silica layers decreases from 3 to 2 as channel aspect-ratio decreases due to the presence of contact-angle boundary condition. **d,** SEM images of packed column of 5 μm silica beads. **e,** Optical microscopic images of assembled silica microsphere structures in rhombus-shape (top left), triangle-shape (top right), square-shape (bottom left) and star-shape (bottom right)

Silica gel microspheres (particle diameter: 3 μm, surface pore diameter: ~ 60Å, see detailed information in supplementary information) which are widely used materials in conventional chromatography have been also self-assembled by using the presented technique. Solid-state silica gel particles were diluted with medium and the droplet of the acquired suspension was applied to the channel. Although silica gel particles shows wider distribution of diameters, the interface between assembled structures and outlet reservoir has been clearly defined and the particles have been assembled tightly in both vertical and horizontal scale as can also be seen from Figure 3.

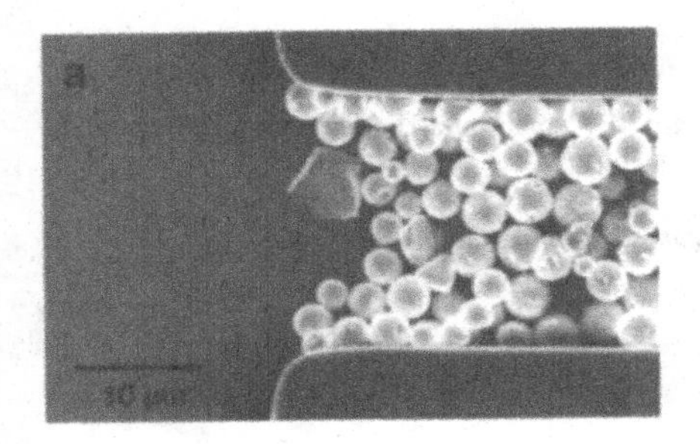

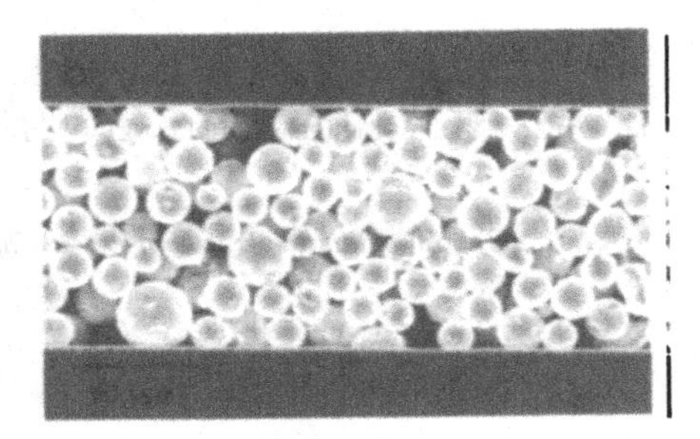

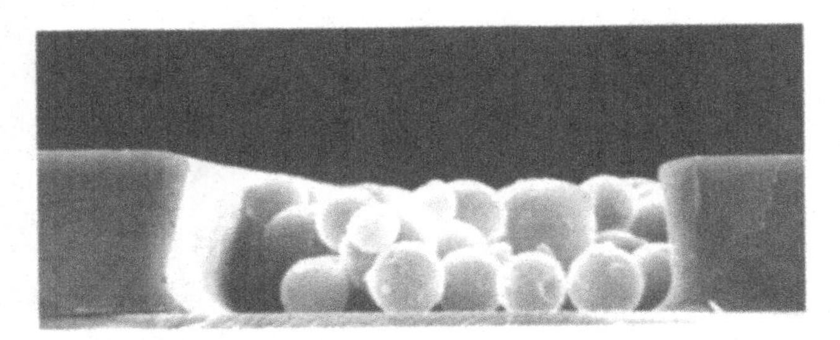

Figure 3 SEM images of assembly of non-uniform micro-scale particles within SU-8 micro-open channel. **a-c**, (a) Interface between channel and outlet (b) Channel Middle (c) Cross section of channel

CONCLUSIONS

We have demonstrated a self-assembly of micro particles by using simple photolithography and surface tension-induced capillary action of micro particle-dispersed suspension. Due to its simplicity and wide applicability to various types of particles and substrate, we expect this presented technique to be widely used in future bio-assay, photonics and semiconductor technologies.

ACKNOWLEDGMENTS

The authors would like to thank Center for Nanoscale Mechatronics and Manufacturing (CNMM), Korea and S. Choi would like to thank Samsung Scholarship Foundation for his graduate fellowship.

REFERENCES

1. Denkov ND, Velev OD, Kralchevsky PA, Ivanov IB, Yoshimura H, Nagayama K. *Nature*, 1993;361:26

2. Joseph Sherma, *Handbook of Thin-Layer Chromatography*, CRC Press, 2003

Advances in Integrating Device Components

Mater. Res. Soc. Symp. Proc. Vol. 1191 © 2009 Materials Research Society 1191-OO08-01

The MAESFLO Device: A Complete Microfluidic Control Systems

Jacques Goulpeau, Vélan Taniga and Charles-André Kieffer
FLUIGENT SA, 29 rue du faubourg Saint Jacques, 75014 PARIS, FRANCE

ABSTRACT

In spite of considerable efforts, flow control in micro-channels remains a challenge owing to the very small ratio of channel/supply-system volumes, as well as the induction of spurious flows by extremely small pressure or geometry changes. We present here a robust and complete system for flow control in complex microchannel network that both monitors and controls all the flow relevant parameters, that is to say flow rate and pressure.

INTRODUCTION

Based on a dynamic control of reservoir pressures at the end of each channel and external thermal flow-sensors, all the parameters are measured with a precision down to 25 μBar and 2nL/min. Thanks to feed back control loop, the MAESFLO can control either the flow rate or the pressure with high stability over long period whatever the microsystem characteristics may be. Compared to classic pumps, a significant increase of stability has been reached as no mechanical parts are involved. Indeed the flow rate is pulse free and is stable down to 0.1% [2] of the full scale. Besides, pressure control enables to achieve short response time (less than hundreds of millisec). The MAESFLO is composed of two elements: The MFCS (MicroFluidic Control System) and the Flowell (flow sensors).

The MAESFLO is thus a unique system to control flow in complex network architecture and can be considered as an alternative to integrated micro-valves using only external equipments. The MAESFLO can indeed stop the flow to nearly zero in one or several branches of a complex microfluidic network while keeping other flows constant. Sequential manipulation of liquids in a definite part of a micro-device is thus possible without expensive and time consuming fabrication processes. It can be highly useful when dealing with washing steps in the case of biological assay for example.

Controlling flow with short response time along with high precision is also a key issue in microfluidic. By combining pressure actuation with flowrate monitoring, short response time are achievable keeping a high precision flow rate. It can be useful for droplet generation and size control, droplet on demand generation, long time living cell perfusion and drug injection for instance.

In this work we will present the benefit to control and monitor both pressure and flow rate with the MAESFLO. A lot of information can be extracted from these simple parameters, as hydraulic resistance, monophasic and biphasic apparent viscosity, the volume and the position of a trapped air bubble and many more. The proof of concept of stop flow control will also be shown with experimental results stressing the advantages of the "virtual micro-valve".

THEORY

Microfluidics flows are characterized by very small Reynolds number that compares inertia energy over viscous energy. Thus simplifications in the Navier-Stokes equation can be made leading to a linear equation called the Stokes Equation:

$$-\frac{\partial p}{\partial x}+\eta\left(\frac{\partial^2 v_x}{\partial x^2}+\frac{\partial^2 v_x}{\partial y^2}+\frac{\partial^2 v_x}{\partial z^2}\right)=0 \qquad (1)$$

With p the pressure, v_i the velocity over the i=x, y, z axis, and η the dynamic viscosity. After derivation of this equation, simple analogy between electricity and hydraulic appears: pressure (P) and flowrate (Q) are equivalent to voltage and intensity. The Ohm law is also verified, defining the hydraulic resistance (R) of a conduct [2]:

$$P = RQ \tag{2}$$

For example, R equals $128\pi L\eta/d^4$ for a tube of diameter d and length L and $12L\eta/h^3w$ for a rectangular channel of height h, width w and length L with h<<w. It becomes easy to derivate relations between pressure and flow rate in complex microfluidic networks. In the following, we consider several situations: a cross shape microdevice and a simple channel with a bubble.

Cross shape microdevice

Figure 1A shows the cross shape channel that is considered in the present section and in the experimental one. Figure 1B shows the modelization given in term of electrical scheme.

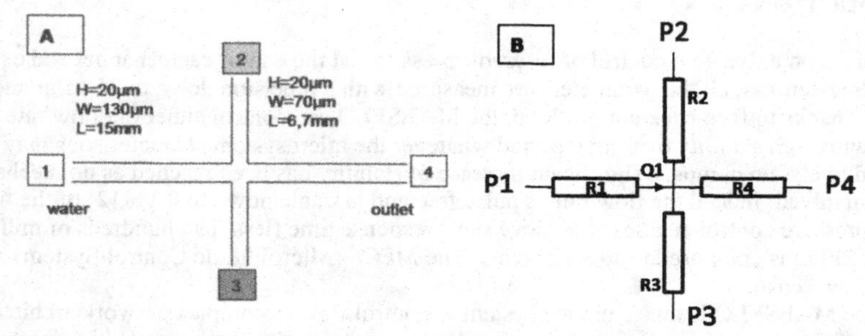

Figure 1: (A) dimension of the cross shape microdevice considered here, (B) the equivalent electrical circuit.

After some calculation, we can express Q1 as a function of P1, P2, P3 and P4:

$$Q1 = \frac{P1}{R1} - \frac{R_{tot}}{R1}\left(\frac{P1}{R1} + \frac{P2}{R2} + \frac{P3}{R3} + \frac{P4}{R4}\right) \tag{3}$$

With $R_{tot} = (\sum 1/Ri)^{-1}$. So by controlling properly the pressure at all the extremity of the microdevice, it is possible to turn the flow rate Q1 to any value, including zero, in a certain range defined by the resistance values and the accessible pressure range.

Simple channel with a bubble

Let's consider a micro-channel characterized by the fact that a air bubble of volume Vb is trapped in the channel (Figure 2A). Its position is defined by the hydraulic resistance before it (R1) and after it (R2). This situation is representative of many current experimental situations in microfluidic, as air bubbles are usually captured in the microsystem when filling it.

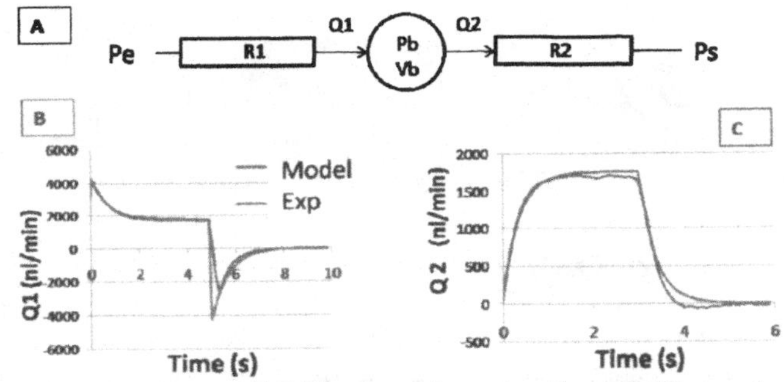

Figure 2: (A) Modelization of the air bubble present in a microchannel, (B) experimental measurement of the flow rate before the air bubble and (C) after the air bubble when modifying Pe. We note a very good concordance between measures and modelization. The bubble volume is equal to 0,44µL.

As we introduce hydraulic resistance, it is possible to introduce hydraulic capacitance [1] that is a very good approximation for modeling the presence of the air bubble. All calculus done, we find:

$$(P_e - P_s) + \tau_1 \frac{dP_e}{dt} = (R_1 + R_2)\left(Q1 + \tau_3 \frac{dQ1}{dt}\right) \quad (4)$$

$$(P_e - P_s) - \tau_2 \frac{dP_s}{dt} = (R_1 + R_2)\left(Q2 + \tau_3 \frac{dQ2}{dt}\right) \quad (5)$$

$$\text{with} \quad \tau_1 = \frac{R_2 V_b}{P_b} \qquad \tau_2 = \frac{R_1 V_b}{P_b} \qquad \tau_3 = \frac{R_1 R_2 V_b}{(R_1 + R_2) P_b} \quad (6)$$

With Pb the initial pressure of the bubble and Vb its volume. We note that the differential equations are different if one considers the flow rate before (Q1) or after (Q2) the bubble. Figure 2 B and C show results of modelization using equation 4 and 5 for a rapid increase (similar to a Heaviside excitation) of pressure Pe.

EXPERIMENT

MFCS (MicroFluidic Control System)

In the nanoliter range, the use of syringe, peristaltic or piston pumps, leads to hysteresis, long equilibration times, irreproducibility and pulsing. Existing fluid handling devices are thus often inadequately adapted to the manipulation of small fluid volumes. Syringe or peristaltic pumps may seem suitable for applications requiring knowledge of the transported fluid volumes. However, actual instantaneous flows in microchannels often have little to do with the pump's setting due to hysteresis and connections compliance. As shown below, the flow rate imposed by a syringe pump strongly depends on the tubing and microsystem characteristics. One may have stable flow rate but with very large equilibrium time, in some case exceeding hours (with elastic tubing) or at the other extremity, short response time can be achieved (with rigid tubing) but important pulsations due to mechanical hysteresis appear (Figure 3).

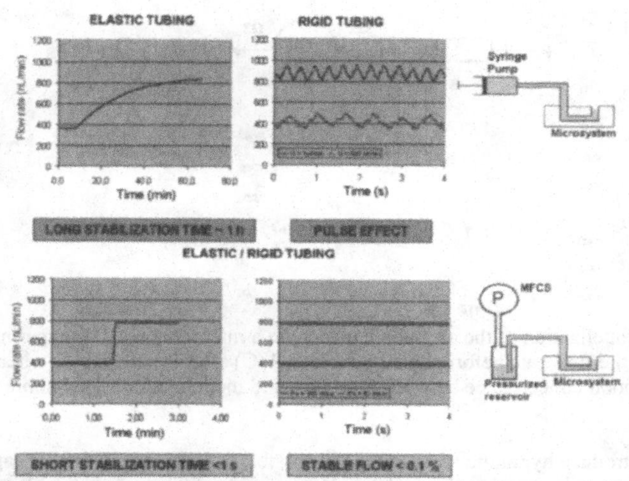

Figure 3: Above, flow rate behavior when using syringe pump with elastic or rigid tubing. Below, the same systems using the MFCS.

Controlling flows with FASTAB™ [2] technology leads to a completely different behavior. The time response is no longer dependent of the compliance of the tubing and the microsystem: as soon as the dynamic pressure is applied, the effect is instantaneous in the whole device. Besides, in many cases, calculating flows in a microfluidic network controlled by the MFCS Flow sequencer is easier and more accurate than using volume-based pumps. Table 1 resumes the main characteristic of the MFCS.

The MAESFLO device

The MAESFLO device is composed of the MFCS and the FLOWELL (Figure 4). The Flowell is composed of 3 independent bi-directional flowmeters ranging from -7 to 7µL/min with a precision of around 2nL/min. The control of the MFCS and the FLOWELL is done with a unique software allowing the control and the monitoring of all the relevant hydraulic parameters in the device. Table 1 resumes the main characteristic of the FLOWELL.

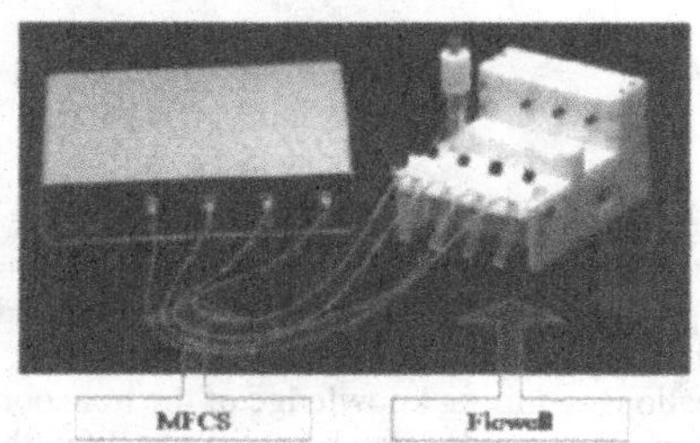

Figure 4: the MAESFLO device is composed of the MFCS and the Flowell. It is composed of 4 to 8 independent pressure channel with a precision of 0.1% and 3 to 6 bi-directional flowmeters ranging from -7 to 7µL/min with a precision of around 2nL/min

MFCS		**FLOWELL**	
Smallest pressure step	25µBar	**Resolution**	2nL/min
Accuracy	1%	**Accuracy**	5,2%
Stability	0.1%	**Precision at 4000nL/min**	0,1%
Independent channels	2 to 8	**Flowrate range**	-7 to 7µL/min

Response / Stabilization time	40ms /1s	Wetted material	Glass, PEEK, PP
Pressure range	0 to 25, 69, 345 or 1000mBar	Max response time	150ms
		Chemical resistance	1M acid and base

Table 1: characteristics of the MFCS and the FLOWELL used for the experiments.

Cross shape microdevice

A microdevice from IBIDI was used for the experiments with the dimensions shown in Figure 1A. A 4 channels 345mBar MFCS was used to control the flows. The liquids were food color dye or DI water previous degassed. The pressures were first set using equation 3 to have Q4 equal to 1µL/min and Q1=Q2=Q3 (Figure 5A). Still following equation 3 for Q2 and Q3, pressures were set in order to reach Q2=0 (figure 5C) and Q3=0 (figure 5B). Only small modification of the calculate pressures were necessary to achieve results shown in figure 5. These small differences between theory and experiments were mainly due to hydrostatic pressure difference induced by non uniform reservoir height.

Simple channel with a bubble

The experiments were done using two flowmeter from a FLOWELL and two pressure control channel from a 4 channel 345mBar MFCS device. An air bubble was introduced in the tube connecting the two flowmeter. As the tube was transparent and of known diameter, the bubble volume (Vb) was estimated by measuring its length. Figure 2B and 2C show example of experimental results. In the model, the resistance R1 and R2, Vb, Pb were calculated or measured. The only fitting parameter was the water viscosity. As shown in Figure 2 and on others experiments (data not shown) the model described above fits the experimental data with a high precision.

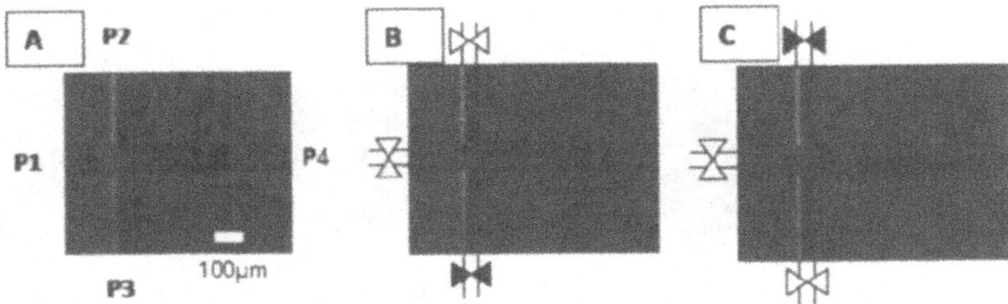

Figure 5: different sets of pressure leading to a co-flow (A), a nearly zero flow rate in the branch number 3 (B) or 2 (C). It is possible to change between two different flow patterns within a fraction of second.

Furthermore, by fitting the experimental data with the solution of Eq. 4 and 5, it is possible to estimate R1, R2 and τ_3. From Eq. 6, it is possible to locate in term of hydraulic resistance and to measure the volume of an air bubble trapped in the device.

CONCLUSIONS

The MAESFLO device is a unique tool to study flows in microdevice. Thanks to high precision and short response time, it is possible to get more precise and even new type of information by controlling and monitoring pressure and flowrate. As shown here, virtual valves are possible thanks to the high precision and stability of the MFCS device. Air bubbles, which are most of the time a difficulty for microfluidic experimenters can now be detected and located. Many others experiments are possible, as viscosity determination for simple or biphasic fluids.

REFERENCES

1. P. Tabeling, "Introduction to microfluidics", Oxford University Press, 24 November 2005.
2. C. Fütterer, N. Minc, V. Bormuth, J.-H. Codarbox, P. Laval, J. Rossier and J.-L. Viovy, Lab Chip, 2004, 4, 351

Mater. Res. Soc. Symp. Proc. Vol. 1191 © 2009 Materials Research Society 1191-OO08-04

Miniaturization of Immunoassays Using Optical Detection With Integrated Amorphous Silicon Photodiodes

A. T. Pereira[1,2], V. Chu[1], D. M. F. Prazeres[2,3] and J. P. Conde[1,3]
[1]INESC Microsistemas e Nanotecnologias and IN- Institute of Nanoscience and Nanotechnology, Rua Alves Redol 9, Lisbon, Portugal
[2]Centro de Engenharia Biológica e Química, IBB – Institute of Biotechnology and Bioengineering, Instituto Superior Técnico, Av. Rovisco Pais, Lisbon, Portugal
[3]Dept. of Chemical and Biological Engineering, Instituto Superior Técnico, Av. Rovisco Pais, Lisbon, Portugal

ABSTRACT

Immunoassays are currently the main analytical technique for quantification of a wide range of analytes of clinical, medical, biotechnological, and environmental significance with high sensitivity and specificity. Miniaturization of immunoassays is achieved using microfluidics coupled with integrated optical detection of the antibody-antigen molecular recognition reaction using thin-film amorphous silicon (a-Si:H) photodiodes. The detection system used consists of an a-Si:H photodiode aligned with a polydimethylsiloxane (PDMS) microchannel. An enzymatic reaction taking place in the microchannel yields a product which is a light-absorbent molecule and hence can be optically detected by the integrated photodiode. Specific antigen-antibody reaction was detected and distinguished from the non-specific reaction.

INTRODUCTION

This work combines a polydimethylsiloxane (PDMS)-based microfluidic sample handling aligned with integrated optical detection based on thin-film amorphous silicon (a-Si:H) photodiodes to achieve a miniaturized immunoassay detection system. An enzymatic reaction taking place in the microchannel yields a product that can be optically detected by the integrated photodiode.

Thin-film silicon photodiodes have high photosensitivity, low dark current and can be deposited at low temperatures (below 250 °C) thus allowing the use of glass and polymer substrates. Their fabrication is an established technology allowing large area fabrication, for example in digital X-ray imagers [1] and the fabrication of a large array of sensors for multiplex measurements. a-Si:H photodiodes have been used as biosensors for chemiluminescent and colorimetric measurements, including immunoassay detection, using reaction volumes in the 10-50 μL range [2-4].

Currently, the method most commonly used for immunoassays is the sandwich enzyme-linked immuno sorbent assay, ELISA, in which enzymes are used as detection labels and are typically carried out in polystyrene microtiter plates [5]. Immunoassays are a multistage, labor-intensive, and time consuming process. Automation of microtiter plate immunoassays can be achieved by the use of complex and bulky robotic systems for fluid manipulation. Miniaturization of immunoassays in a microfluidic system has the potential to provide fast, simple, sensitive, automated, and multiplexed immunoassays, with reduced consumption of sample and reagents and the possibility of bringing the analysis to the point-of-care. Microfluidic techniques allow the manipulation of small quantities (10^{-9} to 10^{-18} L) of fluids in channels with dimensions typically in the range of 10-100 μm [6]. Reactions inside the microchannels will be faster since the diffusion distances of the molecules are much smaller (in the μm range) than in the typical microtiter plates (in the mm range). In addition, the small size greatly reduces the volumes of reagents required thus reducing the cost of the analysis. Development of immunoassays in the microfluidic format started in late 1990s [5]. Further research is required to develop a universal method that can be easily adapted to any antigen-antibody pair, as is now the case with the standard ELISA protocol.

DESCRIPTION OF THE SYSTEM

A model immunoassay in which the antigen is itself an antibody, immunoglobulin G (IgG), was selected. The first step consists in the adsorption of the antigen to the microchannel walls (Figure 1 *(a)*). The number of molecules that are bound to this solid surface is critical for the sensitivity of the reaction and can be adjusted by varying the concentration of antigen in solution and the immobilization conditions, such as exposure time, flow, and surface functionalization. In a microtiter plate assay this step is usually performed overnight at 4 ºC to allow full binding. A surface blocking step is then required usually using bovine serum albumin (BSA) (Figure 1 *(b)*). Next, the antibody specific against the first antibody (Anti-IgG) is allowed to react with the antibody previously immobilized in the microchannel (Figure 1 *(c)*). Each of these steps requires at least one hour when the analysis is carried out in a microtiter plate format. The complete assay, including several washing steps, can take up to 24h. The secondary antibody is labeled with the enzyme horseradish peroxidase (HRP) that reacts with the appropriate substrate returning a colored product that can be quantified by light absorption (Figure 1 *(d)*). The quantity of the colored product is related to the antigen quantity using a calibration curve.

Two main reactions contribute to the signal obtained in the assay: (i) the immobilization of the antigen; and (ii) the antibody-antigen molecular recognition reaction. Two factors are involved in the antibody-antigen reaction: (i) the diffusion of antigens to the channel wall where the antibodies are immobilized, and (ii) the equilibrium of the reaction of formation of the antibody-antigen complex. Since the affinity constant is usually high, varying from 10^5 mol^{-1} to 10^{12} mol^{-1}, the antibody-antigen reaction is expected to be diffusion controlled [7]. Therefore, due to the small dimensions of microchannels, this reaction reaches completion within minutes.

MATERIALS AND METHODS

Photodiode fabrication

200 μm x 200 μm a-Si:H p-i-n photodiodes were microfabricated on glass substrates [3]. Briefly, an Al bottom electrode is deposited on Schott AF45 glass and patterned using wet etching. Next, a p-i-n a-Si:H diode is deposited by rf (13.56 MHz) plasma enhanced chemical vapor deposition (PECVD) n-layer (a-Si:H doped with P), followed by a i-layer (undoped a-Si:H), and a p-layer (a-Si:H doped with B). A diode mesa structure is patterned by photolithography and reactive ion etching (RIE). An insulating layer of silicon nitride (SiN_x) deposited by PECVD is used as a sidewall passivation layer and a via is opened to allow electrical contact between the top transparent conductive oxide (indium-tin-oxide, ITO) and the a-Si:H p-layer. A double layer passivation is deposited, SiN_x followed by SiO_2, both by PECVD. The finished device was then wire bonded to a PCB plate. A top view micrograph of the diode is

shown in Figure 2 *(a)* and a schematic vertical cut of the device is shown in Figure 2 *(b)*.

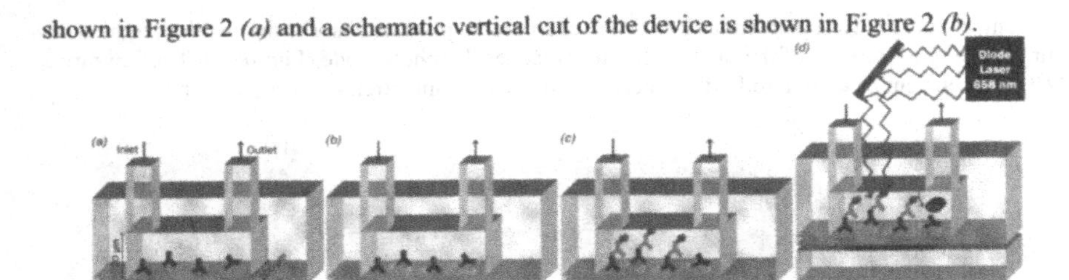

Figure 1 - *(a)* PDMS microchannel scheme with inlet and outlet highlighted. The first antibody is immobilized by adsorption on the channel walls. *(b)* In the second step, the channel surface is blocked with BSA. *(c)* The secondary antibody is inserted in the microchannel and reacts specifically with the first antibody. *(d)* The microchannel is aligned with the photosensor and with the light source. A photocurrent measurement is initiated and the TMB substrate is inserted into the microchannel. In the presence of HRP, TMB is converted into a blue-coloured molecule that absorbs light in the red region of spectrum.

PDMS microfluidic channel fabrication

Microchannels 20 μm high and 200 μm wide were fabricated by soft lithography in polydimethylsiloxane (PDMS) using an SU-8 mold on a Si base substrate. An Al on quartz shadow mask was used for the UV lithography of the SU-8-2015 (Microchem) mold. PDMS (Sylgard 184 – Dow Corning) was prepared by mixing curing agent and base in a 1:10 weight ratio followed by vacuum de-airing. In order to cast the microchannels together with access holes a polymethylmethacrylate (PMMA) mold was fabricated using a CNC milling machine (TAIG Micro Mill from Super tech & Associates). After assembling the SU-8 mold and the PMMA plates (Figure 2 *(c)*), PDMS was injected in the reservoirs and cured for 2 hour at 60 °C. The base of the channel was a 500 μm thick flat PDMS slab that was prepared by spinning the PDMS mixture at 250 rpm for 25 s. The channel was sealed to the base after surface treatment with a corona discharge (Electro-Technic Products).

Adsorption of antibody labeled with HRP

Antibody labeled with HRP was first adsorbed at the microchannel surface. Anti-Goat IgG HRP-labeled (SIGMA) diluted in phosphate buffer (PB) 100 mM, pH 7.4, in concentrations from 0.25 mg/mL (6.25 μM) to 0.00025 mg/mL (6.25 nM) is injected with a flow rate (Q) of 0.5 μL/min for 10 minutes. A syringe pump (New Era Syringe Pump) was used for constant flow injection. Subsequent washing is performed by flowing 15 μL of PB at a $Q = 5$ μL/min for 3 minutes.

For the detection of the enzymatic reaction, the PDMS channel was manually aligned to the top of the photodiode, as shown in Figure 3 *(a)*. A diode laser with a 658 nm wavelength (Power Technology) is used as the light source. The laser light is optically aligned to the channel, and the integrated photodiode. TMB substrate was injected with $Q = 2.5$ μL/min for ~6 min. With the TMB formulation used (TMB liquid substrate system for membranes (SIGMA)), the reaction product precipitates in the place of reaction and accumulates in the microchannels walls as long as there is fresh substrate flow. The accumulated product absorbs the incident light at 658 nm, causing a decrease of the photon flux that reaches the photodiode (Figure 1 *(d)* and Figure 2 *(f)*). The current of the photodiode is measured during the injection using a picoammeter.

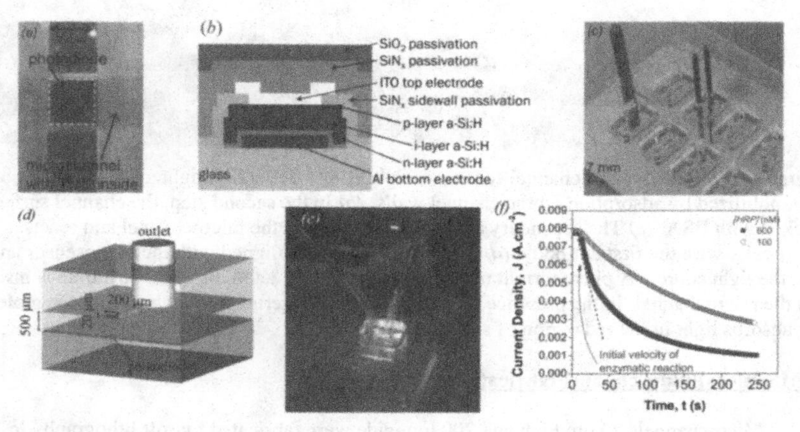

Figure 2 - *(a)* Optical micrograph of the microchannel with TMB product inside overlaying the photodiode. *(b)* Schematic side view of the photodiode. *(c)* Assembly of PMMA plates and silicon base for PDMS casting. The SU-8 mold of the microchannel can be seen underneath the PMMA plates. Pins to define access holes in the PDMS are inserted in one of the reservoirs to show the alignment between the all assemblies. *(d)* Schematic side view of the microchannel, where reactions take place, aligned on top of the photodiode (not to scale). *(e)* Photo showing the PDMS structure on top of the photodiode with red laser light shining through the microchannel. *(f)* Decrease in the photocurrent measured by the a-Si:H photodiode as the enzymatic reaction is initiated.

Antibody/antigen molecular detection and recognition

The primary antibody, Goat IgG (SIGMA) 0.1 mg/ml in carbonate buffer 50 mM, pH 9.6, was injected in the microchannel with Q of 5 µL/min for 1 min and then incubated for 2 h at room temperature (RT) (Figure 1 *(a)*). The washing solution (WS), 0.1% (w/v) BSA (Eurobio) and 0.05 % Tween 20 (SIGMA) in PB, was flowed with Q = 5 µL/min for 3 minutes to wash the channel, and then flowed with Q = 0.5 µL/min for 10 minutes to block the surface and decrease non-specific interactions ((Figure 1 *(b)*). The secondary antibody, Anti-Mouse IgG HRP-labeled or Anti-Goat IgG HRP-labeled (both from SIGMA), diluted 1:60 in PB was then injected in the channel with Q = 0.5 µL/min for 10 minutes ((Figure 1 *(c)*). Finally, the channel was washed by flowing PB with Q = 5 µL/min for 3 minutes. Detection was performed as described above. The total assay time including detection was approximately 2h30 min.

RESULTS AND DISCUSSION

A schematic of the PDMS microchannel aligned with the photodiode and a photo of the

Photodiode optoelectronic characteristics

The dark and photo *J-V* curves of the photodiode are shown in Figure 3 *(a)*. Within the range of light intensities used, the photocurrent measured is proportional to the incident photon flux (Figure 3 *(b)*). The sensor can detect a 425 nm wavelength light with a photon flux down to $\Phi \approx 10^9 \text{ cm}^{-2}.\text{s}^{-1}$.

Adsorption of antibody labeled with HRP

Comparing the absorbance levels for different initial concentrations of HRP-labeled antibody injected and adsorbed inside the microchannel (Figure 3 *(c)*) shows that different levels of adsorbed antibody can be measured and distinguished using this system. The minimum antibody concentration in the solution that can be detected using the adsorption conditions described above is ~ 20 nM.

Antibody/antigen molecular detection and recognition

Immunoassays were performed in the microchannel and the signal was acquired by the integrated photodiode. A total volume $V = 5$ µL and a $Q = 0.5$ µL/min were used for the antigen-antibody recognition reaction. The washing steps were performed with $V = 15$ µL and $Q = 5$ µL/min. The impact of the adsorption of the primary antibody layer on the detection of the seconddary antibody was compared using: (i) the standard conditions used in microplate conditions (incubation overnight at 4 °C); (ii) incubation for 2 h at room temperature; and (iii) injection with V of 5 µL and Q of 0.5 µL/min. The results presented in Figure 3 *(d)* show that conditions (i) and (ii) yield similar results while condition (iii) presented almost no signal. This indicates that the density of the primary antibody depends strongly on the immobilization conditions. Figure 3 *(d)* also shows that the non-specific recognition reaction of the antigen using a different target antibody, anti-mouse IgG-HRP, resulted in low absorbance signal, demonstrating that the washing and blocking steps were effective.

CONCLUSIONS

Immunoassays were performed using a microchannel and integrated detection. The current system allows the reduction of the total assay time from ~24 h to ~2h30m, demonstrating the potential for immunoassay time reduction in microfluidic format. Optimization of the primary antibody surface coverage after washing and blocking is crucial for the sensitivity of the test.

ACKNOWLEDGMENTS

The authors would like to thank J. Bernardo, F. Silva, V. Soares for help in clean room processing and device packaging. Authors thank to A. Pimentel for photodiode design development and to F. Cardoso, J. Loureiro and J. Germano for help in PMMA plates design and fabrication and PCB plate design and fabrication. This work was supported by Fundação para a Ciência e a Tecnologia (FCT) through research projects and Ph.D. grants. V. Chu acknowledges a travel grant from the Gulbenkian Foundation.

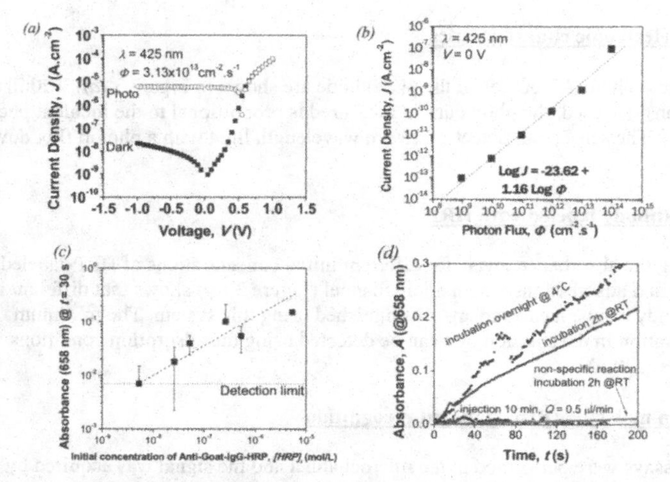

Figure 3 –*(a)* Experimental *J-V* characteristics of the a-Si:H photodiode measured (closed circles) in the dark and (open °circles) under illumination with λ = 425 nm and photon flux Φ = 3.13×10^{13} $cm^{-2}.s^{-1}$. *(b)* Current density of the photodiode sensor, *J*, as a function of the incident photon flux Φ at 425 nm in the absence of an applied bias. The line represents a linear fit to the experimental results *(c)* Absorbance levels measured 30s after the start of TMB injection for adsorbed HRP-labeled antibody injected inside the microchannels in solutions with different concentrations. *(d)* Real time absorbance measured inside the microchannels for three different conditions of adsorption of the antigen Goat IgG. Non-specific reaction was tested using an antibody, Anti-Mouse-IgG HRP, that does not recognize the antigen Goat IgG.

REFERENCES

[1] R. A. Street, *Hydrogenated amorphous silicon*: Cambridge University Press, 1991.

[2] A. C. Pimentel*, et al.*, "Detection of chemiluminescence using an amorphous silicon photodiode," *Ieee Sensors Journal,* vol. 7, pp. 415-416, Mar-Apr 2007.

[3] A. Gouvea*, et al.*, "Colorimetric detection of molecular recognition reactions with an enzyme biolabel using a thin-film amorphous silicon photodiode on a glass substrate," *Sensors and Actuators B-Chemical,* vol. 135, pp. 102-107, Dec 2008.

[4] A. T. Pereira*, et al.*, "Chemiluminescent Detection of Horseradish Peroxidase Using an Integrated Amorphous Silicon Thin Film Photosensor," *IEEE Sensors Journal,* vol. Accepted for publication, 2008.

[5] A. Bange*, et al.*, "Microfluidic immunosensor systems," *Biosensors & Bioelectronics,* vol. 20, pp. 2488-2503, 2005.

[6] H. Parsa*, et al.*, "Effect of volume- and time-based constraints on capture of analytes in microfluidic heterogeneous immunoassays," *Lab on a Chip,* vol. 8, pp. 2062-2070, 2008.

[7] T. G. Henares*, et al.*, "Current development in microfluidic immunosensing chip," *Analytica Chimica Acta,* vol. 611, pp. 17-30, Mar 2008.

Mater. Res. Soc. Symp. Proc. Vol. 1191 © 2009 Materials Research Society 1191-OO08-05

Miniaturized silicon apertures for lipid bilayer reconstitution experiments

Michael Goryll[1] and Nipun Chaplot[1]
[1]Arizona State University, Department of Electrical Engineering,
Tempe, AZ 85287-5706, U.S.A.

ABSTRACT

Ion channels reconstituted into lipid bilayer membranes can be used as a very sensitive and selective platform for high-throughput drug screening applications. In order to employ suspended lipid bilayer membranes for these experiments in form of a "lab-on-a-chip" configuration, a robust and affordable platform is required. In our study, we investigated the feasibility of hosting lipid bilayer membranes across micron-size apertures ranging from 5 µm – 50 µm in silicon. On these substrates, lipid bilayers were formed and characterized concerning their seal resistance, capacitance and breakdown voltage. Seal resistance values of up to 60 GΩ could be achieved repeatedly on these substrates.

INTRODUCTION

Suspended lipid bilayer membranes are an ideal platform to study ion channel behavior, since they closely resemble the natural environment of the ion channels. However, these membranes have to be suspended to allow access to both sides for ion current measurements. The key feature of the membrane support structure is the high seal resistance between the bilayer lipid membrane and the solid support, which has to be on the order of Gigaohms. The commonly used plastic substrates for lipid bilayer suspension are limited when it comes to the miniaturization of the apertures as well as the reproducibility of the apertures. Materials such as PDMS [1] or glass [2] have been investigated as solid supports. Here, silicon microfabrication shows a way toward manufacturing a high density of identical apertures that can easily be combined with microfluidics for a truly integrated microanalysis approach. The process steps needed for silicon microfabrication are well-established and thus readily available and cost effective.

With silicon being an ideal material when it comes to the fabrication, there are several drawbacks which make it unsuitable for bilayer attachment. Silicon dioxide is hydrophilic, thus the attachment of the inner hydrophobic membrane layer to the substrate surface is energetically not favorable. In previous studies we could show that a plasma-polymerized layer of tetrafluoroethylene (C_4F_8) renders the silicon surface hydrophobic and allows for a repeatable formation of bilayer lipid membranes across apertures in silicon.

In earlier work, we chose the aperture size to be 100-150 µm in diameter to be comparable with commercially available substrates [3]. This, however, limits the minimum distance in an array platform. Consequently, a reduction in pore diameter is desirable, in particular since studies on apertures in reflowable Teflon membranes have shown that membranes across apertures on the order of few tens of microns show improved noise characteristics [4].

EXPERIMENT

For our experiments, we used double-sided polished Si (001) wafers with a thickness of 380-420 µm. To enable lipid bilayer thinning, a recess etch was performed to thin down the area around the central aperture. The area for the back-side etch was photolithographically defined. One type of samples underwent an anisotropic etch in KOH, while another design used a deep reactive ion etch (Bosch process) to prepare the recess. In case of the KOH etch the remaining silicon thickness was 30 µm, in case of the deep reactive ion etch the substrate was thinned to 20 µm of remaining silicon. Using back side alignment, bilayer apertures in the range 5 µm and 50 µm were defined photolithographically and etched using the deep reactive ion etch process again. The back side of the wafer was coated with SU-8 of a thickness of 20 µm or 40 µm respectively, with a central area of 100 µm removed around the final aperture. Before dicing the wafer into individual pieces each holding a single aperture, the surface was coated with a plasma-polymerized PTFE layer using the ICP tool. For testing, a single die was mounted into a Teflon holder using silicone gaskets. The Teflon holder has compartments in contact with the silicon surface that hold up to 5 ml of electrolyte solution. 1M KCl buffer, stabilized at pH 7.4 was used for the ion transport measurements. Lipid bilayers were formed using the painting method as described in [5]. Electrical measurements were performed using Ag/AgCl wire electrodes immersed in the baths on either side of the membrane. The electrodes were connected to either an Axopatch 200B or a HEKA EPC-8 patch-clamp amplifier, both operating in resistive feedback mode at a gain setting of 1mV/pA. Membrane breakdown voltages were determined using a Keithley 236 SMU connected directly to the Ag/AgCl electrodes in place of the patch-clamp amplifier.

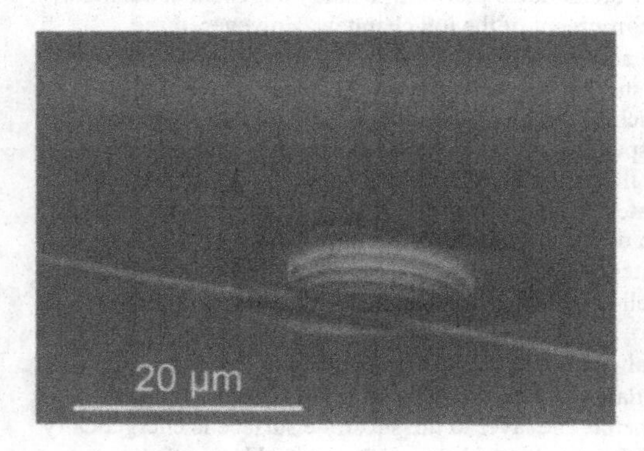

Figure 1: Scanning electron micrograph of an aperture of 15 µm diameter etched in silicon using deep reactive ion etching. After patterning the sample has been oxidized, resulting in an SiO_2 layer of 800 nm thickness. As a final step, 200 nm of plasma-polymerized PTFE were deposited. The diagonal line indicates a scratch in the PTFE layer.

DISCUSSION

The main goal of reducing the aperture size is to accomplish a smaller bilayer area. This should render the bilayer less prone to due to lipid raft fluctuation, thereby increasing bilayer stability. Moreover, a reduction in the bilayer area reduces the bilayer capacitance and therefore the noise associated with the setup. Since the noise of the setup is given by

$$i^2 = i_{th}^2 + i_D^2 + i_e^2 + i_n^2 \, , \qquad (1)$$

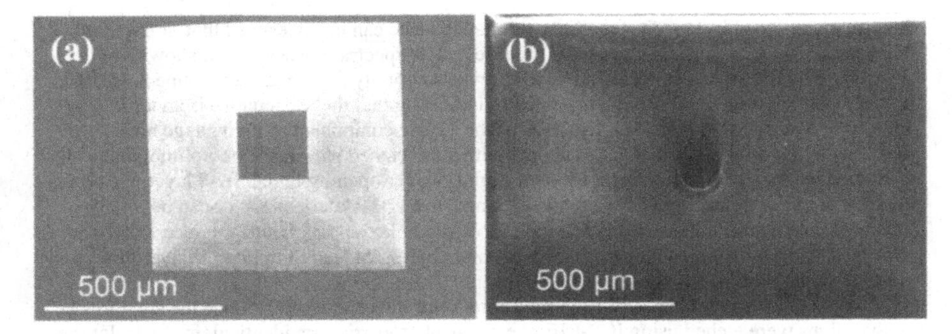

Figure 2: Scanning electron micrographs of the back side etches. The photolithographically defined etch masks had a diameter of 100 µm in either case. Panel (a) demonstrates a recess etch using KOH wet chemistry, resulting in the lateral expansion of the recess, while the reactive ion etch in (b) preserves the original diameter of the etch mask. The shadows in (b) are due to charging effects and the thickness variation of the SU-8 layer.

with the individual contributions being the thermal noise i_{th} of the seal, the dielectric noise of the insulating material i_D, the Miller feedback noise i_e and the current noise of the input transistor i_n. While the thermal noise is white, i.e. does not exhibit frequency dependence, the noise due to the capacitive components increases with frequency [6]. In order to accomplish a large measurement bandwidth, the capacitance has to be reduced as much as possible. In the current setup, the capacitance of the platform has been minimized by using an SU-8 epoxy layer as an additional dielectric besides the SiO_2 as well as a minaturized bilayer aperture.

Using the KOH back-side etch based on an initial patterned structure size of 100 µm, the lateral dimensions of the recess expand to around 740 µm based on the sidewall angle determined by the facet (see Fig. 2). This not only limits the minimum spacing of the bilayer apertures, but also complicates the optical patterning of the SU-8 layer. Due to its viscosity and wetting properties of the silicon surface, frequently a discontinuity in the SU-8 layer develops at the edge of the recess. This opens up the SU-8 layer, leading to an increased contact area and

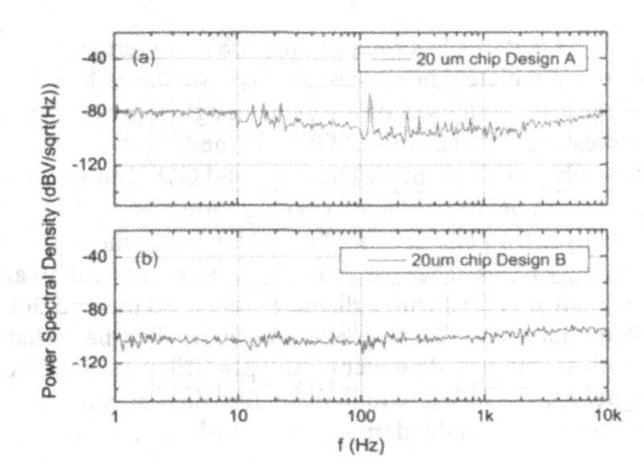

Figure 3: Power spectral density of the noise originating from a lipid bilayer spanning a 20 µm wide aperture in silicon, using (a) KOH back side etch or (b) RIE back side etch. Due to the high capacitance of the design in (a), the noise increases above 1 kHz, limiting the usable bandwidth of the setup.

thus an increased chip capacitance. Since this capacitance can easily exceed that of the lipid bilayer, the noise properties of the setup are affected. A spectral noise analysis shown in Fig. 3, covering the range between 1 Hz and 10 kHz reveals not only an increased baseline noise, but also an increase with higher frequency, clearly indicating that the capacitance is an issue. The higher capacitance also leads to a feedback of the 120Hz component of the voltage noise, resulting in the discrete peak. On the sample with a dry etched recess, the photolithographically patterned recess size is maintained. Using a slightly larger opening in the SU-8 layer, the recess can be completely cleared of SU-8, while a homogeneous thickness of epoxy can be accomplished. Despite the straight sidewalls, no air bubble trapping inside the recess can be observed. Moreover, the cylindrical recess geometry does not lead to lipid multilayer formation nor does it prevent the formation of lipid bilayers.

Lipid bilayer membranes were formed across the etched apertures. Since the apertures in both designs were etched using ICP-RIE, the physical properties are identical, in particular the surface roughness. Thus we expect similar bilayer seal resistance and formation probability. However, bilayers on the samples with the KOH recess etch tend to have a smaller seal resistance and a lower formation probability (no formation at all or rapid degradation of the seal resistance within seconds after the formation). This can be attributed to the properties of the hydrophobic coating. In case of the KOH recess samples, a PTFE layer of 30 nm was deposited and the thickness checked by spectroscopic ellipsometry.

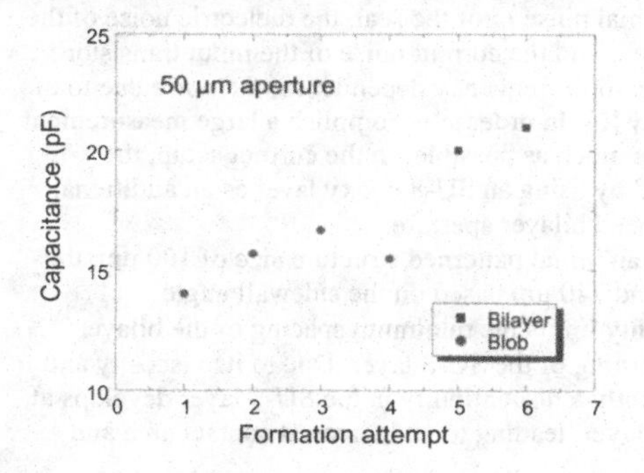

Figure 4: Lipid bilayer capacitance values for subsequent formation attempts. After approx. 3-4 attempts, bilayers form and the capacitance increases to a value larger than 20 pF.

Although previous measurements indicate that this layer has hydrophobic properties, a layer of this thickness does not seem to form a completely homogeneous layer on the etched inner surface or does not smoothen out the intrinsic roughness of the etch enough to provide an optimum surface for bilayer attachment. Increasing the deposited PTFE thickness to about 200 nm results in repeatable seal formation with seal resistance values up to 60 GΩ. Multiple subsequent bilayer formation on the same aperture provides statistics on the formation probability. Typically 10 to 20 subsequent bilayer formations using the lipid painting method could be accomplished without having to reapply lipids. The seal formation was successful on all pore sizes and the seal resistance did not degrade on pores with a diameter below 30 µm and did degrade only slightly on apertures with 50 µm diameter. This degradation was smaller than what has been observed on the earlier samples with an aperture diameter of 150 µm. Bilayers could be distinguished from multilayers by their higher capacitance due to their smaller thickness in the central area. This indicates that the bilayer capacitance is the dominating contribution to the overall capacitance and thus the chip capacitance has been reduced to a value that contributes only negligible to the input capacitance. Measurements artificially clogging the aperture show a chip capacitance of 4-6 pF. Thus a capacitance measurement is sufficient to determine the success of a bilayer formation (cf. Fig. 4).

Using the voltage-gated OmpF ion channel of E. coli we were able to demonstrate that the membranes formed across the apertures allow demonstrate ion channel reconstitution into these lipid bilayers. Fig. 5 shows the channel activity, indicating the physiological behavior of the channel in the particular membrane.

CONCLUSIONS

We were able to fabricate apertures in silicon ranging from 5 μm – 50 μm using dry reactive ion etching. Lipid bilayer formation was successful on all sizes, with the bilayers being most stable for aperture sizes below 50 μm. Seal resistances in the tens of Gigaohms could be accomplished on a regular basis and the formation of a lipid bilayer concluded from a capacitance measurement. The noise of the platform critically depends on the homogeneity of the SU-8 layer, which is why the dry reactive ion etching of the back side recess is preferred over a wet chemical etch using KOH. OmpF ion channels could be reconstituted in membranes formed across these miniaturized apertures.

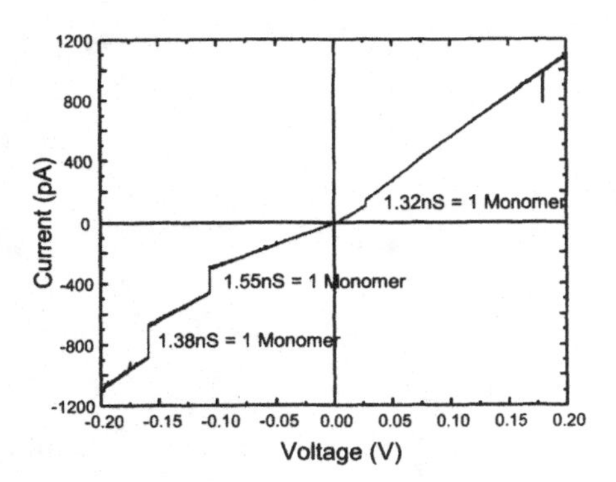

Figure 5: Current-voltage characteristic of a single OmpF ion channel inserted in a lipid bilayer membrane across a 50 μm wide aperture etched in silicon. The back side etch was performed using reactive ion etching. The measurement indicates physiological gating, thus the geometry of the pore allows for the thinning of the bilayer lipid membrane.

REFERENCES

[1] K.G. Klemic, J.F. Klemic, M.A. Reed, and F.J. Sigworth, *Micromolded PDMS planar electrode allows patch clamp electrical recordings from cells.* Biosensors & Bioelectronics **17**(6-7), 597-604 (2002).

[2] N. Fertig, Ch. Meyer, R.H. Blick, Ch. Trautmann, and J.C. Behrends, *Microstructured Glass Chip for Ion Channel Electrophysiology.* Physical Review E **64**, R40901-1-R40901-4 (2002).

[3] S.J. Wilk, L. Petrossian, M. Goryll, T.J. Thornton, S.M. Goodnick, J.M. Tang, and R.S. Eisenberg, *Integrated electrodes on a silicon based ion channel measurement platform.* Biosensors & Bioelectronics **23**, 183-190 (2007).

[4] M. Mayer, J.K. Kriebel, M.T. Tosteson, and G.M. Whitesides, *Microfabricated Teflon Membranes for Low-Noise Recordings of Ion Channels in Planar Lipid Bilayers.* Biophysical Journal **85**, 2684-2695 (2003).

[5] M.D. Mager, and N.A. Melosh, *Lipid bilayer deposition and patterning via air bubble collapse.* Langmuir **23**(18), 9369-9377 (2007).

[6] R.A. Levis, and J.L. Rae, *Constructing a patch clamp setup.* Methods Enzymol. **207**, 14-66 (1992).

Porous Materials in Labs on a Chip

Mater. Res. Soc. Symp. Proc. Vol. 1191 © 2009 Materials Research Society 1191-OO09-02

Efficient Nanoporous Silicon Membranes for Integrated Microfluidic Separation and Sensing Systems

Nazar Ileri[1,2], Sonia E. Létant[2], Jerald Britten[2], Hoang Nguyen[2], Cindy Larson[2], Saleem Zaidi[3], Ahmet Palazoglu[1], Roland Faller[1], Joseph W. Tringe[2] and Pieter Stroeve[1]

[1]University of California, Davis, Davis, CA 95616, U.S.A.
[2]Lawrence Livermore National Laboratory, Livermore, CA 94550, U.S.A.
[3]Gratings, Inc., Albuquerque, NM 87107, U.S.A.

ABSTRACT

Nanoporous devices constitute emerging platforms for selective molecule separation and sensing, with great potential for high throughput and economy in manufacturing and operation. Acting as mass transfer diodes similar to a solid-state device based on electron conduction, conical pores are shown to have superior performance characteristics compared to traditional cylindrical pores. Such phenomena, however, remain to be exploited for molecular separation. Here we present performance results from silicon membranes created by a new synthesis technique based on interferometric lithography. This method creates millimeter sized planar arrays of uniformly tapered nanopores in silicon with pore diameter 100 nm or smaller, ideally-suited for integration into a multi-scale microfluidic processing system. Molecular transport properties of these devices are compared against state-of-the-art polycarbonate track etched (PCTE) membranes. Mass transfer rates of up to fifteen-fold greater than commercial sieve technology are obtained. Complementary results from molecular dynamics simulations on molecular transport are reported.

INTRODUCTION

Transport of biomolecules through nanopores is crucial in many biotechnological and biophysical processes [1]. Advances in protein screening, organic and inorganic molecular development, and sensing of toxins/viruses have accelerated the requirement for membranes with uniform pore size and a large dynamic range of biomolecule size selection (~1 nm to 1 μm). However, the standard lithographic processes limit the production of membranes with nanometer-scale pore sizes over 1-100's of mm^2 areas needed by most applications. The majority of commercial membranes are made of organic polymers and fabricated by non-lithographic methods [2]. For example, PCTE membranes, produced by ion-track etching through polycarbonate films, have pore sizes in the range ~10 nm to ~μm. Although the scalability of these membranes is good, uniformity and flow rates are limited to ±20% and <0.1 $mL/min/cm^2$ (for 10 nm diameter pores), respectively [3]. Other porous filters such as anodic aluminum oxide (AAO) and mesoporous silica have been created through anodic etching of Al [4, 5], and sol-gel processing [6], respectively. The pore dimensions are variable in the range ~ 30-400 nm for anodic alumina and ~ 2-20 nm in sol-gel films. Membranes made of zeolites, on the other hand, have ultra-uniform pores, but only in the relatively narrow range of ~0.3-3 nm [7].

The most common state-of-the-art membranes, PCTE and AAO sieves, have cylindrical pores which exhibit relatively low transport rates and are prone to clogging. Sieves with conical pores, by contrast, have been reported to exhibit dramatically higher transport rates compared to the analogous cylindrical pore membranes [8], but these membranes have to date been limited by pore size uniformity and mechanical strength issues, for example. Here we present initial performance results of novel silicon membranes created by a new synthesis technique based on interferometric lithography. Silicon has a number of advantages as a filter material over the polymers, including good mechanical, thermal and chemical stability, potential for well-controlled pore size and ready sterilization. Porous silicon membranes reported here are created by interferometric lithography (IL), a process which enables rapid definition of nanoscale features over large areas through interference of two laser beams. With appropriate processing steps, millimeter-sized planar arrays of uniformly tapered nanopores can be created in silicon with pore diameter 100 nm or smaller. For this work, porous Si membranes were fabricated, and pyridine diffusion experiments conducted to test the membranes' efficiency against state-of-the-art commercial sieves.

THEORY AND EXPERIMENT

For the fabrication of nanopore arrays, 4 inch, P-doped silicon-on-insulator SOI (100) wafers were used, with a top silicon layer thickness of ~380 nm. Si_3N_4 was deposited on both sides of the wafer to serve as Si etch mask. An antireflection coating (ARC) was used to suppress the standing wave formation resulting from substrate back-reflection. Next, a positive resist was deposited on top of the ARC and baked. The pattern was generated by large-area IL using a 413 nm Kr-ion laser as a light source. Pattern definition and ARC removal were followed by an electron-beam deposited Cr layer, and then lift-off of the resist film. Cr served as a mask for etching silicon nitride. After etching Si_3N_4 down to the Si layer using an ion beam, a larger-scale pattern on the back side was created photolithographically to define the freestanding membrane areas, and the handle wafer was removed. Finally, pits were etched on the front Si layer, and the buried oxide layer was removed by vapor-phase HF etching.

To characterize the pores, the wafers were imaged after each processing step using atomic force microscopy (AFM) and scanning electron microscopy (SEM). AFM was conducted on a Nanoscope V5 and SEM with FEI 430 NanoSem Electron Beam Lithography System. To obtain statistical data on pore uniformity, images were analyzed with Scion Image Beta 4.0.3. Mass transfer experiments were conducted using pyridine (HPLC grade, Aldrich Inc.) molecules. Commercial PCTE membranes with pore diameter in the ~ 100nm range (Poretics, Inc.) were used as received together with fabricated Si membranes. The membrane separated the diffusion cell into two compartments, reservoir and sink. The reservoir contained 1 mM aqueous solution of pyridine (30 mL), and the sink contained DI water (30 mL) without pyridine. The change of pyridine concentration in the sink compartment was measured by UV-visible light spectrophotometry (Varian, Cary 3) at 255nm.

For the molecular dynamics (MD) simulations, a solvent free coarse-grained model was used where proteins were treated as colloidal particles. The model was implemented by means of the ESPResSo (Extensible Simulation Package for Research on Soft matter) molecular dynamics package [9]. Length, energy and time were measured in units of the diameter of protein beads (σ), the cohesion potential depth between protein beads (ε), and $\tau = \sigma\,(m/\,\varepsilon)^{1/2}$ (where m is the bead mass), respectively. The temperature was chosen as $k_BT = 1.0\varepsilon$. Since the thermal energy unit at room temperature is $k_BT = 4.1 \times 10^{-21}$ J $\approx$ 0.6 kcal/mol (at T=298 K), ε is thus fixed at 0.6 kcal/mol. Finally, by using $\tau = \sigma\,(m/\varepsilon)^{1/2}$ and the known value for the protein mass, τ was found to be 5.6ps. The time step was set to $\delta_t = 0.005\ \tau$. 30600 protein particles were placed in a $100\sigma \times 100\sigma \times 200\sigma$ simulation box subject to periodic boundary conditions in all three dimensions. Temperature control was achieved by a dissipative particle dynamics (DPD) thermostat which ensures correct hydrodynamics. The friction constant was determined according to $\Gamma = \tau^{-1}$. The interactions between the filter and the protein beads as well as protein-protein beads were set to a purely repulsive Weeks-Chandler-Anderson potential [10], while there was no interaction between filter beads. Visualization of simulation snapshots was performed with VMD [11].

DISCUSSION

Molecular Modeling

Three different pore geometries, cylindrical, conical, and pyramidal (Fig. 1), were studied with large scale ESPResSo simulations. Higher diffusion rates were obtained with tapered geometries compared to the cylindrical geometry at short simulation times (τ=1500, results not shown). This result was expected as the wall hindrance is much higher in cylindrical geometry in comparison with tapered pore geometry. It supports our experimental approach, based on tapered pore geometries.

Figure 1. Simulation configurations for cylindrical, conical, pyramidal pores.

Membrane Fabrication and Characterization

The initial pattern on the resist was defined by IL. IL exposures can be strongly affected by conditions such as laser power. However, in general, the technique is capable of creating large-area fields of features with good uniformity. The IL uniformity can be compared, for example, to the uniformity of pores in PCTE membranes (see Table 1, Fig. 2(a, b)). Fig. 2(a) is a scanning electron microscope (SEM) image of a PCTE membrane, and Fig. 2(b) is an SEM image of an 833 nm period 2D photoresist pattern created by IL after exposing and developing. After defining the pattern and etching the anti-reflective coating down to silicon nitride, a 75 nm Cr layer was deposited onto the surface and the resist layer was lifted-off (Fig. 2(c)). The silicon nitride served as a second etch mask for KOH etching on the front and back sides of the wafer since it exhibits improved mechanical stability and etch selectivity compared to the grown oxide layer.

The final features were created in silicon by etching after Si_3N_4 etch and handle wafer removal. Aqueous KOH solution etches Si in a highly anisotropic manner, and hence creates near-perfect inverted pyramids (shown in Fig. 2(d, e)), but conical pores (used in preliminary diffusion experiments reported here) can also be used created in Si by more isotropic dry or wet etch processes. In the final processing step, buried oxide was removed with vapor-phase HF. Fig. 2(e) demonstrates top-down view of a complete membrane wafer.

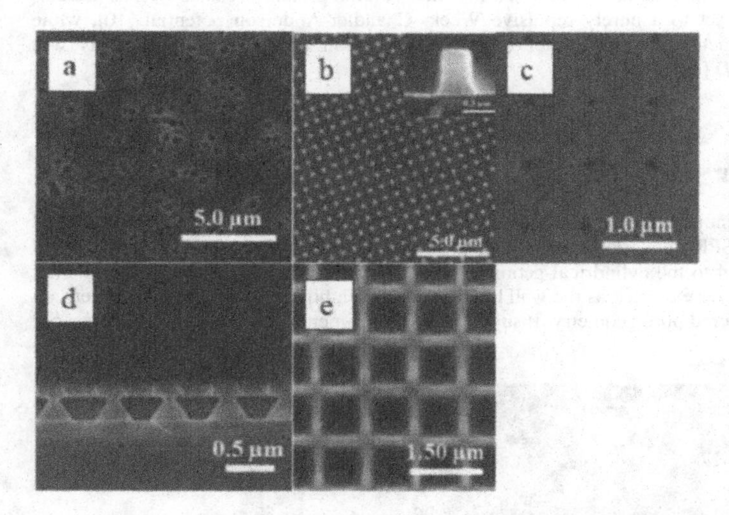

Figure 2. Scanning electron microscope images of (a) polycarbonate track etched (PCTE) membrane, (b) photoresist defined by interferometric lithography, (c) resist lift-off after 75 nm Cr deposition, (d) V-grooves after front side KOH etching, (e) processed membrane wafer.

Table 1. Image analysis results comparing silicon membrane uniformity versus PCTE.

Pore/Membrane	PCTE	Pattern on resist	Silicon Filter
Diameter (nm)	326	≤345	100
Std. Dev. (nm)	249	69	48
Std. Error	0.76	0.20	0.48

Transport Experiments

The efficiency of the fabricated filters in comparison to PCTE was investigated through a set of preliminary experiments. Pyridine was used in these experiments because the molecule diameter is significantly smaller (~0.5 nm) than the pore size (~100 nm), therefore allowing pure diffusion effect to be investigated. Table 2 shows mass transfer results for Si and PCTE membranes. Fifteen-fold higher fluxes were obtained for Si membranes compared to state-of-the-art PCTE membranes with ~100 nm pore diameter.

Table 2. Pyridine flux through nanoporous Si membranes versus commercially, available PCTE membranes.

	PCTE	Si Membrane
Pore size (nm)	100	≤100
Membrane thickness (nm)	6000	280
Pore density (pores/cm^2)	6×10^8	3×10^8
Pyridine flux (mole/cm^2.s)	5.5×10^{-10}	9.9×10^{-9}

CONCLUSIONS

This study demonstrates the performance results of novel silicon membranes created by a new synthesis technique based on interferometric lithography. Uniform pores with diameter 100 nm or smaller were fabricated in silicon over millimeter-sized areas. Pyridine diffusion through silicon membranes was compared against the state-of-the-art polycarbonate track etched (PCTE) membranes. Pyridine transported through conical pores was found to be fifteen fold greater than through commercial filters. These observations were supported by molecular dynamics simulations showing higher diffusion rates for tapered pores compared to cylindrical pores. These newly created nano-filters could be integrated into a multi-scale microfluidic processing system and are expected to be useful for improving the selectivity, speed and efficiency of molecular separations for biomedical and biodefense applications.

ACKNOWLEDGMENTS

We would like to thank Harold Levie, Matthew Hoopes and Rodney Balhorn for assistance. This work was partially supported the University of California System wide Biotechnology Research & Education Program GREAT Training grant 2007-03. Parts of this work were performed under the auspices of the U.S. Department of Energy by Lawrence Livermore National Laboratory under Contract DE-AC52-07NA27344.

REFERENCES

1. C. Dekker, *Nature Nanotechnology*, vol. 2, pp. 209, 2007.

2. W. Chu, R. Chin, T. Huen, M. Ferrari, *J. Microelectromech. Syst.*, vol. 8, no. 1, pp. 34, 1999.
3. Sterlitech, *"Sterlitech Polycarbonate Track Etch (PCTE) Membranes"*, http://buyglassmicrofiber.com/.
4. A. P. Li, F. Muller, A. Birner, K. Nielsch, and U. Gosele, *J. Appl. Phys.*, vol. 84, pp. 6023, 1998.
5. H. Masuda and K. Fukuda, *Science*, vol. 268, pp. 1466, 1995.
6. W. Chen, W. Cai, L. Zhang, G. Wang, and L. Zhang, *J. Coll. Interf. Sci.*, vol. 238, pp. 291, 2001.
7. I. Lenntech, *"Regenerative adsorption zeolite"*, http://www.lenntech.com/.
8. N. Li, S. Yu, C.C. Harrell, C. R. Martin, *Anal. Chem.*, vol.76, pp. 2025, 2004.
9. H.J. Limbach, A. Arnold, B. A. Mann, *C. Holm, Comput. Phys. Commun.*, vol 174, pp. 702, 2006.
10. J. D. Weeks, D. Chandler, H. C. *Anderson, J. Chem. Phys.*, vol. 54, pp. 5237, 1971.
11. W. Humphery, A. Dalke, K. Schulten, *J. Mol. Graph.*, vol. 14, pp. 33, 1996.

Mater. Res. Soc. Symp. Proc. Vol. 1191 © 2009 Materials Research Society 1191-OO09-04

Direct FIB fabrication and integration of "single nanopore devices" for the manipulation of macromolecules

B. Schiedt[1,2], L. Auvray[2], L. Bacri[2], A.-L. Biance[3], A. Madouri[1], E. Bourhis[1], G. Patriarche[1], J. Pelta[2,5], R. Jede[4] and J. Gierak[1]
[1]LPN/CNRS, Route de Nozay, F-91460 Marcoussis, France
[2]MPI, Université d'Évry Val d'Essonne, Bd. François Mitterrand, F-91025 Évry Cedex, France
[3]LPMCN, UMR5586, Univ. Claude Bernard Lyon 1, 43 Boulevard du 11 Novembre 1918 F 69622 Villeurbanne, France
[4]RAITH GmbH, Hauert 18, Technologiepark, D-44227 Dortmund, Germany
[5]Université de Cergy-Pontoise, F-95302 Cergy-Pontoise, France

ABSTRACT

Here we propose to detail an innovative FIB instrumental approach and processing methodologies we have developed for sub-10 nm nanopore fabrication. The main advantage of our method is first to allow direct fabrication of nanopores in relatively large quantities with an excellent reproducibility. Second our approach offers the possibility to further process or functionalize the vicinity of each pore on the same scale keeping the required deep sub-10 nm scale positioning and patterning accuracy.

We will summarise the optimisation efforts we have conducted aiming at fabricating thin (10-100 nm thick) and high quality dielectric films to be used as a template for the nanopore fabrication, and at performing efficient and controlled FIB nanoengraving of such a delicate media.

Finally, we will describe the method we have developed for integrating these "single nanopore devices" in electrophoresis experiments and our preliminary measurements.

INTRODUCTION

Since several years, nanopores in thin solid state membranes, individual or as arrays, have found a growing number of applications for example as stencils or masks to grow or depose nanostructures, or to fabricate single molecule electronic detectors or sensors [1]. The latter is probably one of the most prominent among them and consists in using the membrane as a dividing wall in an electrolytic cell and drawing charged molecules by an electric field through the pore. The resulting current blockage signal reveals information about the passing molecule so that e.g. DNA or proteins can be manipulated and studied at a single molecule level in native conditions. While there is a lot of work going on in this field using biological pores such as α-hemolysin, some tasks which are e.g. implicating higher temperature or aggressive chemicals require the more robust solid state pores. For the realisation of nanopores the main difficulty arises from the observation that standard lithographic techniques are not applicable to the realisation of nanometre-sized holes within free standing films. One the one hand there is the question of the resist coating process that is difficult to combine with nanometre-sized film thicknesses. On the other hand existing final etching / transfer processes remain difficult to control reproducibly at the sub-10 nm scale. Therefore a number of different approaches have been developed in recent years to produce such nanometric pores. These include: (i) sculpting methods, where a larger hole is shrunk using ion or electron beam irradiations at low fluences, (ii) direct drilling with focused electron (TEM) or ion beams as well as lithographic methods and ion track etching. For an overview over current fabrication techniques see e.g. [2]. Considerable progress has been made in producing smaller and smaller pores, but with most of the existing methods, each sample has to be produced almost individually and manually, making them only available to very specialized teams or laboratories.

After analyzing this, we decided then to develop a specific approach aiming at bridging this first gap and allowing further functionalities to be integrated at the nanopore scale, which is important for the considered applications such as e.g. rapid DNA sequencing [3].

Following this we came to the point that what is needed first is a considerably larger scale production capability based on an integrated approach where a plurality of identical devices can be produced in an automated way. With our approach, we are able to write the same pattern (the nanopore itself and a mark to retrieve it) in a batch process allowing to position and engrave one single nanopore on each of the ~200 membranes present in a 2" wafer. For this strategy we use a high resolution FIB instrument developed in our laboratory in combination with an automated high precision sample positioning stage and writing software. We will show hereafter that our nanopore device fabrication approach owns the potential to turn the artificial single nanopore device from a precious research object to something closer to a consumable for biologists or biophysicists.

Existing methods are capable of defining single nanopores, but they do not allow subsequent nano-functionalization or -addressing of the pore, because the tool which was used for the pore fabrication is not able to write locally at the required resolution (pore final diameter). Therefore their final applicability remains limited. Thus it is clear that a direct lithographic tool having the capability to write, i.e. to engrave or depose, must be preferred, and this is the approach we have selected for fabricating a "single nanopore device".

It is obviously important to allow the nanopore localization via nanometre accuracy reference marks in the immediate vicinity of the nanopore. Such references allow the nanopore localization for monitoring its characteristics, using for example the TEM technique, and subsequent realignment for complementary processes.

Finally thanks to these characteristics we were capable to integrate the so produced nanopore samples into a commercially available setup and have observed current-voltage characteristics and single molecule translocations through our fabricated "single nanopore devices" (Figure 1).

EXPERIMENTS AND METHODS

Focused ion beam instrument

We have developed a FIB system dedicated to nanofabrication [4]. This instrument combines the advantages of a very high resolution FIB system, having a deep sub-10 nm predicted patterning resolution, with the accuracy of a high precision laser interferometer stage (2 nm steps) and a high speed digital pattern generator.

The FIB column was designed using a high-performance optical architecture we tested in high-resolution FIB systems [5]. In our present ultra-high resolution optics mode, the maximum ion energy is set to 35 keV. Particle focusing is achieved via two asymmetric lenses working in the decelerating mode and in infinite magnification conditions. In this column the deflection plates are located between the lenses, allowing a reduction of the final lens working distance (WD), a necessary condition to achieve a strong demagnification of the virtual source size (~30 nm). Then the source magnification d_G is smaller than 4 nm. The disks of least confusion for chromatic (d_{ch}) and spherical aberration (d_{sp}) have respective values of ~3 nm and around 0.03 nm.

Using the usual summation method in quadrature, $d^2 = d_G^2 + d_{ch}^2 + d_{sp}^2$, the optimum value of the spot diameter d is found to be about 5 nm (FWHM).

The performance of this high-resolution ion optics column was calculated for various configurations using state-of-the-art theory and modelling software developed and verified experimentally [6]. One basic reason explaining such a record FIB patterning resolution level, down to the sub-5 nm, is that the probe current remains sufficiently high to perform reproducible and controlled nano-fabrication experiments. For low acceptance angles and low Liquid Metal Ion Source (LMIS) emission currents ion-induced shot-noise and statistical fluctuations are observed and are the main limits for FIB ultra-high resolution. In our case the operating conditions applied to our specific LMIS, the on-axis angular intensity (number of particles / unit solid angle) could be increased by a factor of 4 [7]. For the same probe size the final current remains higher (5 to 8 pA for the above mentioned ion probe characteristics). This interesting feature does not simply allow a reduction in exposure time and improved selectivity. It also offers an efficient tuning of both the focusing, stigmatism and the deposited fluence of the ion probe, by using the Ion Microscopy mode during the initial calibration sequence.

Moreover as described elsewhere [8], FIB irradiation of thin membranes allows to replicate very easily the incident FIB probe profile and therefore to control probe resolution, beam profile and residual astigmatism with an extreme sensitivity. Therefore nanopore engraving tests were very early carried out leading to a very precise "beam shaping" methodology and this experience was fully applied in this work.

"Single nanopore device" fabrication

Freestanding membranes of SiC and Si_xN_y (various stoichiometries) were fabricated in different thicknesses down to 20 nm. The process consists of thermally oxidizing 2" (100) silicon wafers, sputtering SiC or Si_xN_y on one side and finally adjusting the stress of this layer by thermal annealing to about 800 MPa. The wafer was then patterned by UV lithography in a way to yield the maximum number of samples possible (Figure 1a), each containing one freestanding membrane (Figure 1a inset). The size of the individual samples was chosen to be small enough (i.e. < 3 mm) to fit easily into standard TEM sample holders for further imaging or analysis. Etching was performed in an aqueous solution of 5% of TMAH, 0.4 % ammonium peroxodisulphate at 80 °C. The width of the lines separating the individual samples was chosen such that after completion of the etching the membranes still hold together but can easily be separated (Figure 1a). In a final step the oxide layers on the back of the wafer as well as on the membrane are removed by Ammonium Fluoride solution.

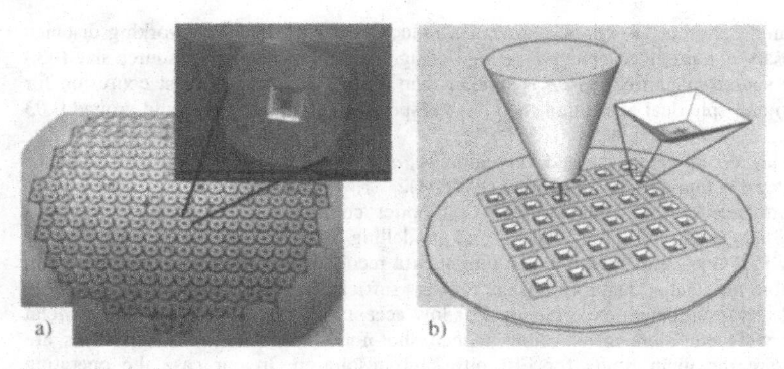

Figure 1. a) Two inch wafer containing 284 membranes. Inset: view of cleaved "nanodevice" (size 3 mm in diameter). b) Scheme of FIB batch patterning process of the complete wafer. Inset: Single nanopore device.

Si_xN_y membranes of 50 nm thickness were also purchased from Protochips, Inc. and used for FIB patterning as received.

Batch patterning of "single nanopore devices"

After having shown the capability of our FIB instrument to directly drill nanopores as small as 3 nm in diameter [6] we aimed at increasing production capability. For this purpose, the same GDSII file that has been used for the fabrication of the optical lithography mask was exploited by the pattern generating software of the FIB. Alignment was performed by locating manually three of the larger membranes and using their centers as references. Then, a pattern consisting of a low dose irradiation mark (Figure 3a) and the nanopore itself was attributed to the center of each membrane and the milling task was completed unattended (Figure 1b). The material of the membranes used for this experiment was 50 nm thick SiC, chosen because of our long experience with it. We aimed at producing pores around 50 nm in diameter. For this, a calibration array consisting of lines of varying doses (similar to the one shown in Figure 2a) was patterned on one of the membranes to choose the right dose. Inspection by optical microscopy after completion of the irradiation showed that all of the membranes had been patterned with the desired pattern. SEM microscopy of 50 of the pores revealed a mean diameter of the pores of 50 ± 9 nm.

This variability is due to slight variations in the membrane thickness over the whole wafer as well as local variations of the microscopic structure of the membrane. If pores are made with the same ion dose close to each other (in one array or in different arrays on the same membrane), they show very reproducible results (Figure 2b).

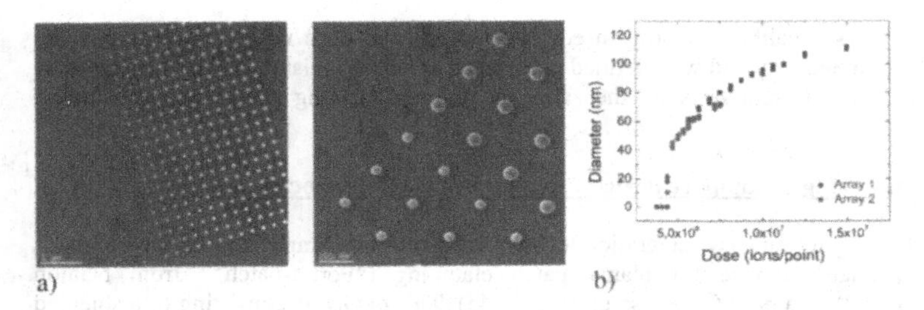

Figure 2. a) Array with columns of different doses in a 50 nm thick Si_xN_y membrane. The dose is increasing from left to right. b) Diameter vs. dose for two arrays made on the same membrane on different days.

As it is possible to engrave sub-10 nm holes, experiments to produce even smaller pores in the same large quantity are currently under preparation. Two routes are currently explored:

Since the size vs. dose curve gets steeper for smaller sizes (Figure 2b) and the imaging facility available in our FIB is inferior to that of a TEM, it will be advisable to do the dose calibration on one of the membranes, then remove this and do a TEM inspection to choose the right dose.

To overcome the limitations of the calibration imposed by local variations in membrane structure, it is planned to implement a system employing a back face ion detector as used in [9], in which the opening of the pore is detected by measuring the transverse current.

Pore characteristics monitoring

Using the low dose alignment marks (Figure 3a) that we patterned in the vicinity of the membrane, the nanopores can easily be found in TEM for high resolution imaging. We found that pores with perfectly round geometry can be made in different materials. Figure 3 shows a selection of pores of different sizes made in Si_xN_y membranes of 50 nm thickness. The black dots are resulting from gallium implantation.

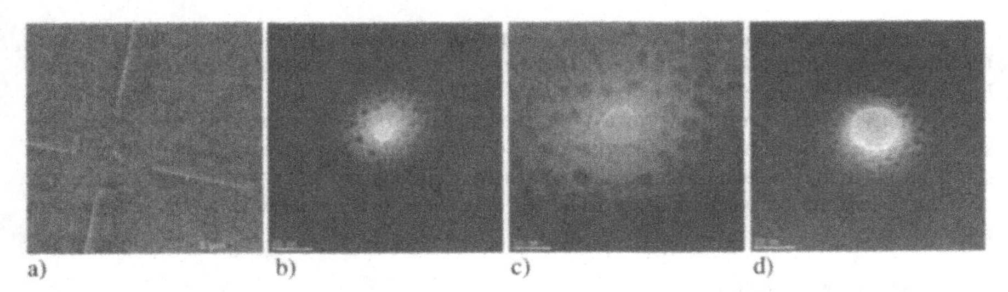

Figure 3. a) TEM image of a pore aligned at the centre of a low dose (~10^2 ions/pt) pattern b-d) TEM images of pores of different sizes in 50 nm thick Si_xN_y drilled with a dose ~10^6 ions/pt

Figure 2a shows a calibration array used to determine the dose needed to perforate the membrane. The homogeneity and well defined pore shape are clearly visible. Two of such arrays have been made on different days on the same membrane, showing a high reproducibility. (Figure 2b).

Integration of the "single nanopore device" in a commercial electrochemical cell

We were looking for an easy integration technique for our membranes and decided to use a commercially available device for planar patch clamping ("Port-a-patch®" from Nanion Technologies GmbH) as a base for our setup (Figure 4a). This device is consisting of a shielded Axon 200B headstage, whose output is connected to a vertical Ag/AgCl electrode, mounted into a cylinder containing an external screw thread, onto which plastic cap, the so called "chip" can be screwed. This chip contains a hole of 1.8 mm diameter in the center, onto which a thin glass plates with a central hole of 1, 10 or 50 µm diameter is normally glued, onto which e.g. a cell can be aspirated. Without this glass, our 3 mm large membrane can be easily mounted onto the hole in the plastic. For this we used a fast curing two component silicone (KwickCast®, World Precision Instruments, Inc.) (Figure 4c,d). Then drops of electrolyte are placed on either side and a second electrode is closing the circuit from the top (Figure 4b).

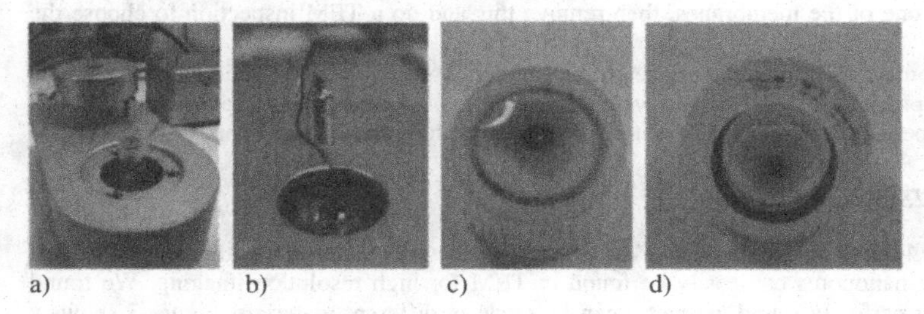

a) b) c) d)

Figure 4. "Port-a-Patch" with upper electrode a) open b) connected. c) and d) membrane glued on plastic chip.

If needed, the membrane can be easily removed from the chip without damage and e.g. further imaged in TEM. This simple setup was used for the measurements shown below, however a disadvantage is the relatively large surface of the sample that is exposed to the electrolyte solution, contributing to a high capacity and possibly increasing the noise. This surface can be reduced by gluing the membrane on one of the glass covered chips with e.g. a 50 µm hole. Since the position of the nanopore is easily visible in an optical microscope via the low dose marks, it can be positioned directly over the hole. Measurements with this setup are in preparation.

Electrolytic measurements

As a final surface treatment membranes have been either subjected to an oxygen plasma from both sides for 30 s – 1 min from each side after mounting to the setup or immersed in

Piranha solution (30 vol% H_2O_2/H_2SO_4, 1:1) for 30 min, to enable wetting of the nanopores by the electrolyte solution.

Two types of macromolecules to be detected have been added to the electrolyte solutions. Phage λ – DNA has been purchased from Sigma and used as received. The multi-domain protein fibronectin was purified from cryoprecipitated plasma. The purity of the preparation was determined equal to 96.6 ± 1.2%.

RESULTS AND DISCUSSION

Performance illustration of "nanopore device"

For the first test of our final device we used pores fabricated in 50 nm thick SiN membranes, since we have found that SiC is less suited due to its non-perfect isolating properties.

Figure 5 shows an I-V curve and some selected DNA and protein current blockade events through a pore in 50 nm thick Si_xN_y. Measurements were performed in 250 mM KCl, 10 mM Tris, 0.5 mM EDTA, pH 7.25 filtered at 2 kHz and sampled at 50 kHz. The I-V curve is linear as for all of our pores. We added lambda DNA (48.5 kbp) as well as fibronectin -a large protein from the extracellular matrix- to the negative side of the membrane. Due to their large size these molecules are ideal to employ in our relatively large pores. Since both molecules are negatively charged, they are drawn by the electric field through the pore. In both cases current blockage events appeared (Figure 5b, c). However, while in the case of the DNA multilevel blockages were observed, which indicate that the DNA blocks the pore in different confirmations, this was not observed for the protein fibronectin. In our experimental conditions, the protein is a flexible string of 56 globules. The hydrodynamic radius of the fibronectin is 11.5 nm and the radius of a globule is 2.5 nm [10]. The current blockade durations should be associated to protein translocation events and are in agreement with the diffusion coefficient of the fibronectin. Each conformation of the protein is similar.

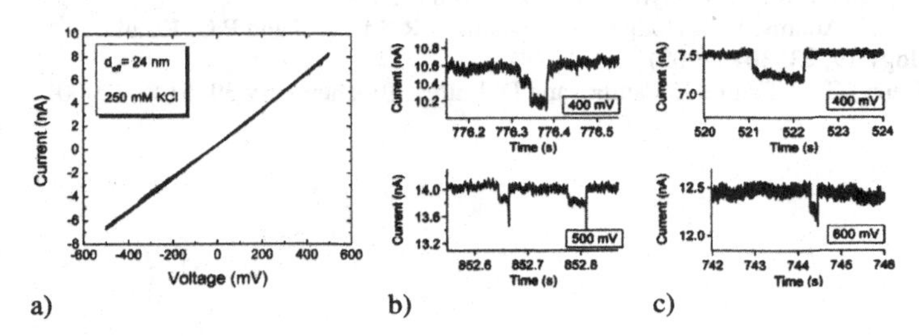

Figure 5. Measurements done with nanopore in 50 nm thick Si_xN_y, dimensions 21 x 27 nm in 250 mM KCl, 10 mM Tris, 0.5 mM EDTA, pH 7.25 a) Linear I-V curve b) blockade events measured at 6.5 nM λ DNA at 400, 500 mV c) blockade events after addition of fibronectin to a final concentration of 0.545 mg/ml.

CONCLUSIONS

The final advantage of our FIB processing technique is to allow the realization of a large number of identical membranes within a single processing batch, yielding the advantage of combining the advanced FIB processing technique together with parallel fabrication such as conventional lithography and etching processes. Such an efficient combination demonstrates that FIB processing can be put in line with the advanced technologies thus allowing new fabrication routes to be open.

ACKNOWLEDGMENTS

This work was partially supported by the ANR "Active nanopore" project and by SESAME contract n°1377, the Région Ile de France and the Conseil Général de l'Essonne.

REFERENCES

1. A.-L. Biance, J. Gierak, E. Bourhis, A. Madouri, X. Lafosse, G. Patriarche, G. Oukhaled, C. Ulysse, J.-C. Galas, Y. Chen, L. Auvray, Microelectronic Engineering 83, 1474 (2006)
2. K. Healy, B.Schiedt, A.P. Morrison, Nanomedicine 2, 875 (2007)
3. D. Branton et al., Nature Biotechnology 26, 1146 (2008)
4. Nano-FIB project. EC GROWTH - G5RD-CT2000-00344 See also ftp://ftp.cordis.europa.eu/pub/nmp/docs/nanofib_info2.pdf
5. J. Gierak, A. Septier and C. Vieu, Nucl. Instrum. Meth. A 427, 91 (1999)
6. J. Gierak, A. Madouri, A.L. Biance, E. Bourhis, G. Patriarche, C. Ulysse, D. Lucot, X. Lafosse, L. Auvray, L. Bruchhaus and R. Jede, Microelectronic Engineering 84, 779 (2007)
7. J. Gierak, Semicond. Sci. Technol. 24, 23 (2009)
8. J. Gierak et al., Mater. Res. Soc. Symp. Proc. Vol. 1089 (2008)
9. N. Patterson. D.P. Adams, V.C. Hodges, M.J. Vasile, J.R. Michael and P.G. Kotula, Nanotechnology 19, 235304 (2008)
10. J. Pelta, H. Berry, G. C. Fadda, E. Pauthe, and D. Lairez, Biochemistry 39, 5146, (2000)

Sensing and Detection on Chip – Molecular Level

Mater. Res. Soc. Symp. Proc. Vol. 1191 © 2009 Materials Research Society 1191-OO10-02

Inkjet Printing of Enzymes for Glucose Biosensors

C. Cook, T. Wang, B. Derby
Manchester Materials Science Centre, Manchester University, Grosvenor Street, Manchester, M1 7HS

ABSTRACT

Drop on demand inkjet printing is a potential method for depositing enzymes onto electrodes for sensor applications. This technology offers drop sizes in the region of picolitres and allows a production rate up to 200 mm/s. This enables not only a more rapid method of device prototyping but also a method for possible miniaturization of the sensors themselves. However, previous work [1] has indicated that inkjet printing may cause a drop in the retained activity of the enzyme.

Here we assess the criticality of this drop in activity and how it may have been influenced by changes to the protein structure during printing. The enzyme used is glucose oxidase and the test methods include; protein analysis, in the form of analytical ultra-centrifugation and circular dichroism, scanning electron microscopy, atomic force microscopy and phase contrast microscopy, to analyse the surface topology of the electrodes and contact angle analysis, to assess the degree of spreading and the interactions between the drops and the electrode surface.

With glucose oxidase there is no change in the conformation, structure or hydrodynamic radius of the protein after printing. The analysis of the electrode surface shows a relatively smooth surface that is made up of individual graphite flakes laid down by a screen printing method. When contact angle and spreading analysis is carried out it demonstrates reliability in the printing process as well as a drop in the sessile volume of the drop in conjunction with a growth in the base diameter of the drop as expected. It also demonstrates a fairly quick rate of evaporation of the drop. Upon the addition of surfactants to the solution the spreading is seen to be more extensive in relation to the surfactant concentration, although some initial reduction in experienced at low concentrations which may be due to the absorption into the carbon surface.

INTRODUCTION

Current production techniques for glucose sensors involve screen printing of the carbon electrode, enzyme (glucose oxidase) and mediator (ferricyanide). This has limitations in terms of the amount of enzyme that can be deposited on the electrode surface and therefore the volume of blood sample that is required to attain a relative reading of blood glucose. The current trend for the blood glucose sensing market is to look towards smaller sampling lancets or lancets with multiple analytes [2]. Inkjet printing may be a relevant technology for this application with its ability to deposit picolitre amounts of fluid rapidly, accurately and with excellent repeatability.

However, previous work carried out on inkjet printing of proteins has reported that damage to the proteins is encountered upon printing, demonstrated by a drop in the retained activity or fluorescence of the protein [3-6]. Therefore, this work has a two fold purpose: to clarify what damage is occurring, if any, upon printing and to rectify this; and to produce a working solution and method for the deposition of the solutions or analytes onto the electrode using inkjet drop on demand printing technology. Thus, this paper will attempt to provide an overview of the printed proteins structure and conformation in comparison to that of a non printed control as well as to provide some surface analysis of the electrodes. The spreading of the solutions on the carbon will then also be analysed over time so that a suitable concentration of surfactant can be added to produce an even spread over the electrode surface and therefore an even deposition of the enzyme for maximal accuracy in its electrochemical properties.

EXPERIMENTAL

Analytical Ultra-Centrifugation (AUC) and Circular Dichroism (CD) analysis

AUC and CD analysis was carried out in the faculty of life sciences (University of Manchester, United Kingdom) at a concentration of 1mgml^{-1} of Glucose Oxidase in Phosphate Buffered Saline.

Scanning Electron Microscopy (SEM)

SEM analysis was carried out by removing a small portion of the Carbon electrode (supplied by Applied Enzyme Technology, Monmouth House, Pontypool, Wales, U.K.) from the electrode sheet and fixing it in the SEM before finding a good image and focus. All scales are given on the images. Images were taken for the electrode with a 1mm cross section and all magnifications and details about the values used are given in each image.

Atomic Force Microscopy (AFM)

Atomic Force Microscopy (AFM) uses a cantilever to maintain a point at fractions of a nanometer distance from the surface of whatever is being measured by reacting to the forces applied as it scans the surface of the test material. This enables for a highly accurate topographical analysis of the specimen, demonstrating features of a few nanometers in height and giving a very high resolution image. The AFM was used in scanning mode over an area of approximately 80um in length.

The electrode was again prepared by removing a small section of the carbon black electrode from the electrode sheet and preparing it in the usual manner for measurement. Again all necessary scales are given on the images. This procedure was repeated for the 1mm diameter electrode.

Phase Contrast Microscopy (PCM)

Phase Contrast Microscopy is a technique that uses interference lines on the sample and the contrast they provide to distinguish areas of different heights and therefore give an overall representation of the surface roughness and the topology of the samples. The difference between this method and the AFM method is that this allows for a wider area of the sample to be scanned as well as for a larger difference in surface roughness to be detected. The areas examined of each electrode were approximately 600um x 200um.

The electrode was placed under the microscope for analysis and the appropriate zoom used in order to detect a reasonable image. All necessary scales are given on the images and the procedure was completed for the 1mm diameter electrodes.

Contact angle analysis

The contact angle analysis was carried out using a camera placed parallel to the material to be imaged so that the surface of the image was flat and horizontal across the camera lens. The sample solution could then be applied to the material using the printing method and a Microfab head (60V dwell voltage with a 3us fall, dwell and rise time as well as 999 drops per trigger at a frequency of 5000Hz) and a series of 250 pictures taken at a spacing of 0.4 seconds (Microfab, Plano, Texas, USA). This allowed for analysis of spreading over a total time of 2 minutes. On board software was then used to measure the angle associated with the solution in the images where it contacted the material. Analysis was then carried out on the degree of spreading across all the samples using a 5% Glucose Oxidase in PBS solution and a number of different concentrations of surfactant (Triton X-100, Sigma Aldrich, UK) to detect any changes in the drop property.

RESULTS AND DISCUSSION

Previous work has demonstrated that when a protein is printed using piezoelectric inkjet technology, a drop in the retained activity or fluorescence is experienced [3-6]. In order to proceed with this work, it was first necessary to analyse the protein after printing, both in terms of its structure and conformation. Figure 1 shows the results attained using AUC and CD:

a)

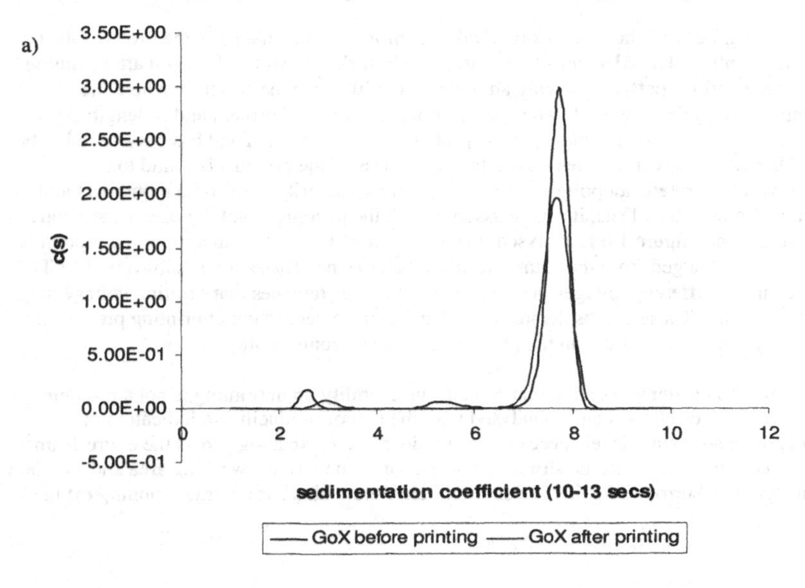

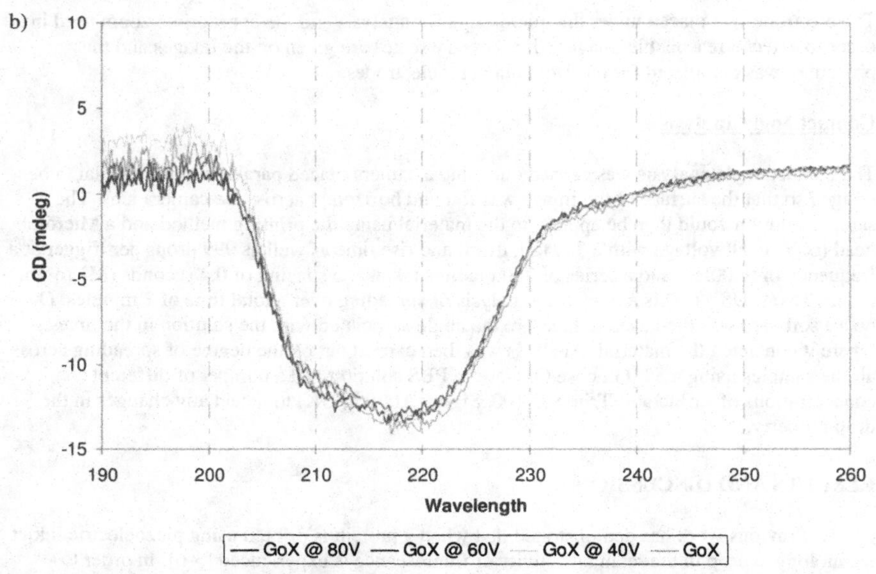

Figure 1. a) AUC analysis of Glucose Oxidase b) CD analysis of Glucose Oxidase printed at different voltages.

Figure 1 demonstrates that there is no structural or conformational change in the protein Glucose Oxidase after printing. The AUC results (Figure 1 a) show the enzyme before and after printing with the sedimentation coefficient giving an indication of the enzyme structure where the Sedimentation Coefficient = f + Rl (where f= friction, R= radius of protein and l= length if protein). Using this equation the radius of the protein can be calculated and has been found to be unchanged between the two samples. Also, the molar mass of the proteins is found to be unchanged with the protein adopting a dimmer form in the majority with some monomers and agglomerate also detected. Thus, it can be assumed that the printing is not altering the structure of the enzyme. When figure 1 b is analysed it can be seen that the conformation of the protein is also remaining unchanged from the natural form of the enzyme. The samples shown in the CD were printed using different voltages in order to detect any differences that printing voltage may have on the samples. These results demonstrate that the piezoelectric inkjet printing process is not having any detrimental effect on the protein structure or conformation.

With the above results demonstrating that protein survivability upon printing is not a problem, the properties of the electrode can be analysed with the aim of producing an indication of spreading and the surfactant levels necessary to produce even spreading across the entire 1mm^2 area of the electrode. The electrode structure consists of a small 1mm^2 working area and a carbon track running off it. Surrounding this is a semi-circular line with a carbon track running off the end of this which is the reference electrode. These are screen printed onto a polyester sheet with approximately 0.5mm between the two electrodes. Therefore it is of great importance to ensure that any spreading of the solutions is prevented from proceeding over the boundaries of the working electrode as this would cause interference with the reference electrode and therefore incorrect electrochemical properties. Figure 2 gives an indication of the surface properties of the 1mm electrode:

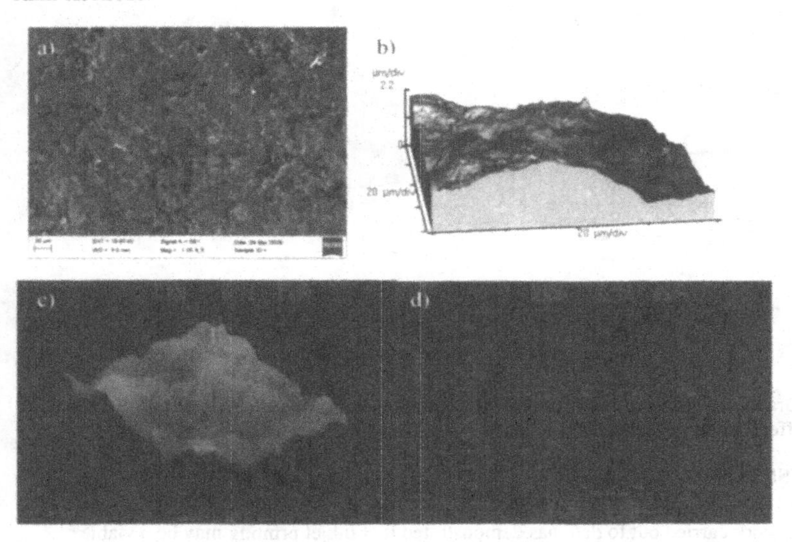

Figure 2. a) SEM image of the 1mm carbon electrode b) AFM image of the 1mm electrode c) PCM surface representation d) PCM graph showing heights of the surface in x and y directions.

Figure 2 shows four different representations of the surface of the 1mm electrode, with the SEM image showing an exact image of the surface and the PCM and AFM showing representations based on the data extracted from the relevant forms of analysis. All of the images show a relatively smooth surface, with the carbon approximately 15um above the polyester substrate. The PV value, which gives an indication of the height of the lowest point to the highest point, is 6.86um and the Rq value, which gives an indication of the overall surface roughness, is 1.22um. Both of these values reinforce the picture taken in the SEM of a smooth surface which is relatively non-porous.

The sample was then analysed using the contact angle method explained above, to assess the spreading of the solution over the sample and the percentage of surfactant required to give the correct degree of spreading. Figure 3 shows the results achieved. As can be seen from figure 3, a small concentration of surfactant causes the contact angle to increase, and there to be less spreading. This is theorized to be due to the absorption of the drop into the electrode more rapidly because of the addition of very small concentrations of surfactants. When a concentration of 0.05% is reached it causes the contact angle to decrease and spreading to occur very rapidly. This is the highest concentration of surfactant that can be used before spreading occurs onto the polyester and the reference electrode. At this concentration, the spreading is so rapid that beyond 40 seconds the contact angle is too shallow for the software to detect. This concentration also allows for maximum spreading without interference of the reference electrode.

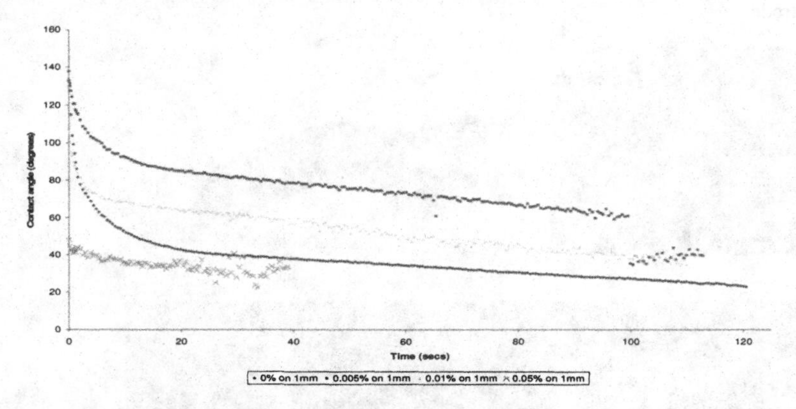

Figure 3. Graph showing the interactions of the solutions with the surface in terms of contact angle at different percentages of surfactants.

CONCLUSIONS

The work carried out to date has demonstrated that inkjet printing may be a viable technology for the deposition of small volumes of protein onto a carbon surface for the sensing market. It is possible to deposit liquid volumes in the range of picolitres, or a number of drops to the range of nanolitres onto a relatively small area. It has also been possible to formulate a solution with the correct concentrations of surfactant and sugars (supplied by AET, Pontypool, UK) present to ameliorate any loss of activity encountered through printing, although this has been proven to not be due to any changes in conformation or structure, and to enable the correct degree of spreading across the electrode surface.

ACKNOWLEDGMENTS

Thanks to Xaar Technologies (Xaar plc, Cambridge, UK), Applied Enzyme Technology (AET ltd, Pontypool, UK), Ellis Technologies (Newark, UK), Oxford Biomaterials (Oxford, UK) and to the Technology Strategy Board (TSB, UK government funding organization, UK) for their continued support of the project both in terms of financial support and mentoring support. Thanks must also go to the Engineering and Physical Sciences Research Council (EPSRC, science funding council, UK) for their continued financial support of the project.

REFERENCES

[1] Nishioka, Markey and Holloway, Protein Damage in Drop-on-Demand printheads, J. Am. Chem. Soc., 126, 16320-16321, 2004.

[2] J. D. Newman, A. P. F. Turner, *Biosensors and Bioelectronics,* **20,** 2435, (2005)

[3] J. G. Mohanty, J. S. Jaffe, E. S. Schulman and D. G. Raible, *J.f Immunol. Meth.,* **202**, 133, (1997).

[4] G. M. Nishioka, A. A. Market and C.K. Holloway, *J..Amer. Chem. Soc.,* **126,** 16320, (2004).

[5] L. Setti, C. Piana, S. Bonazzi, B. Ballarin, D. Frascaro, A. Fraleoni-Morgera and S. Giuliani, *Analytical Letters,* **37**, 1559, (2004).

[6] R. Saunders, J. Gough, N. Reis and B. Derby, in "Architecture and Application of Biomaterials and Biomolecular Materials", Editors: J.Y. Wong, A.L. Plant, C.E. Schmidt, L.Shea, A.J. Coury and C.S. Chen, *Mater. Res. Soc. Symp. Proc.* Vol. EXS-1, F6.3.1 (2003).

Mater. Res. Soc. Symp. Proc. Vol. 1191 © 2009 Materials Research Society 1191-OO10-04

Study of Biosensors Based on Fe Nanowires

YANG Hao and CHEN Yu-quan
State Specialized Laboratory of Biomedical Sensors, Department of Biomedical Engineering, Zhejiang University, Hangzhou 310027, China

ABSTRACT

Anodic aluminum oxide (AAO) template was prepared by using anodizing voltage step-decreasing method after two-step oxide method. Based on AAO template, Fe nanowires arrays were electrochemically deposited. Fe nanowires were coated by chitosan. Fe nanowires/chitosan was synthesized by glutaraldehyde as cross-linking reagent. By crosslinking α-human chorionic gonadotropin (α-HCG), biological probes with Fe nanowires/chitosan/antibody were prepared. An easy operating, easy taking and rapid reacting magnetic detecting system was developed after optimizing the geometry parameters of detect coil. Different concentration samples with 1, 2 and 5 g(Fe)/L were detected. The results show that the sensitivity of system is 0.2 g(Fe)/L and can be improve better.

INTRODUCTION

In the last decade, nanowires and other 1D nanostructures have emerged as potential components suitable to be integrated in devices specially designed for sensing applications such as chemical sensing, biosensing or photodetection. Their use in solid-state sensing technologies has been demonstrated to be very promising due to the novel properties and functionalities derived from their high surface-to-volume ratio, controlled surface interactions, efficient nanoscale transduction mechanisms, and quantum confinement effects[1-3].

EXPERIMENT

Materials and instruments

Aluminum foil (purity, 99.999%, thickness, 0.5 mm) was obtained from Beijing Mengtai Technology and Development Center, China. Chitosan (deacetylation degree >90%) was obtained from Sinopharm Chemical Reagent Co, Ltd., China. Mouse anti-human α-HCG (9.6 g l^{-1}) was obtained from Shanghai Genering Biotech Co, Ltd., China. The other reagents used were analytical reagents and were not purified before use. Deionized water was used in all experiments.

Testing instruments: ultraviolet spectrophotometer, CARY 100, was purchased from Varian, USA. Dispensing system, XYZ3000, was purchased from BioDot, USA. Superconducting quantum interference device (SQUID), MPMS XL-5, was purchased from Quantum Design, USA. Field-emission scanning electron microscope (FSEM), SIRION-100, was purchased from FEI, USA. Transmission electron microscope (TEM), JEM-1200EX, was purchased from JEOL, Japan.

Preparation of Fe nanowires

After suitable cutting, electrochemical polishing, and washing, aluminum foil was dipped in 0.3 M oxalic acid solution with ice bathing, and anode oxidized at 40 V DC for 1 h. It was then soaked in a mixture of phosphoric acid and chromic acid for 1 h, and the same anode oxidation was repeated. After decreasing the anodic oxidation voltage step by step, the AAO template was obtained.

The electrochemical deposition solution was made of $FeSO_4 \cdot 7H_2O$ (100 g l^{-1}), $(NH_4)_2SO_4$ (15 g l^{-1}), $MgSO_4$ (30 g l^{-1}), ascorbic acid (1 g l^{-1}), and glycerin (2 ml l^{-1}). AAO template was dipped in it and the pH was set at 3. Electrochemically deposited at 15 V 50 Hz AC for 5 min.

After removing the aluminum substrate of the template by saturation $SnCl_4$, the AAO template was dissolved by NaOH solution. It was washed with deionized water and enriched by magnetism repeatedly until the pH was 7.0. After supersonic, equable solution of Fe nanowires was obtained.

Preparation of Fe nanowires/chitosan

0.01 g Fe nanowires, 2 ml liquid paraffin, and 0.1 ml span-80 were added in a 1-ml chitosan/acetic acid solution (10 g l^{-1}), supersonic dispersed for 30 min. 0.15 ml glutaraldehyde solution (25%) was added, and shaken slowly at room temperature for 4 h. It was washed with petroleum ether, acetone, and deionized water in turn and enriched by magnetism repeatedly. Then, it was dispersed in 10 ml phosphate buffer solution with pH 7.0, and maintained at 4 °C.

Preparation of biological probes with Fe nanowires/chitosan/α-HCG

The pH of Fe nanowires/chitosan solution was set at 7.2, α-HCG antibody was added, and the concentration was adjusted to 0.1 g l^{-1}, and shaken slowly at room temperature for 24 h. Then, 1 ml BSA (5%) was added, and shaken slowly at room temperature for 6 h. It was washed with phosphate buffer solution and enriched by magnetism repeatedly. It was then rinsed in 5 ml phosphate buffer solution with pH 7.2, and maintained at 4 °C.

Experiment of biological probes with Fe nanowires/chitosan/α-HCG detecting

Frequency mixing is a well-known technique for extracting nonlinear characteristics of materials or of electronic devices. Based on frequency mixing at the nonlinear magnetization curve of superparamagnets, a novel detection technique for magnetic nano-material was recently developed[4].

Upon magnetic excitation at two distinct frequencies f_1 and f_2 incident on the sample, the response signal generated at a frequency representing a linear combination mf_1+nf_2 is detected. The appearance of these components is highly specific to the nonlinearity of the magnetization curve of the particles[5].

The experimental setup of our magnetic detection system is depicted in figure 1.

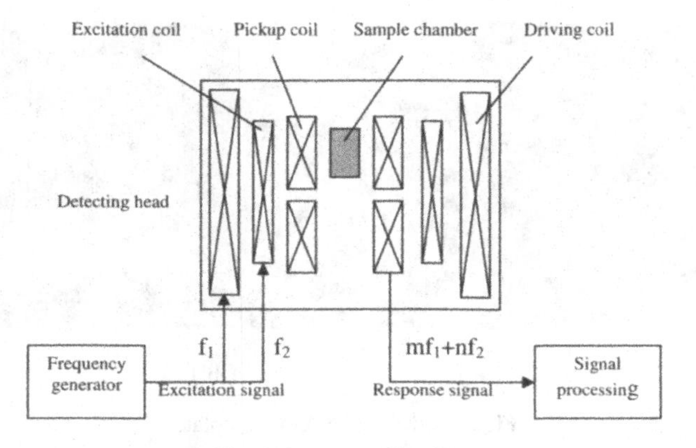

Figure 1. diagram of immunosensor detecting system based on magnetic nanowires

After optimized, the coils were wound with 0.12 mm wire. The parameters of pickup coil were chosen as follows: inner radius is 9 mm, width is 8 mm, height is 1.44 mm and winding number is 800. The parameters of excitation coil which produced an excitation field of 1.5mT were chosen as follows: inner radius is 12 mm, width is 60 mm, height is 0.12 mm and winding number is 500. The parameters of driving coil which produced a driving field of 4.5mT were chosen as follows: inner radius is 11 mm, width is 50 mm, height is 0.12 mm and winding number is 400.

Different concentration samples with 1, 2 and 5 g(Fe)/L were detected by our magnetic detection system.

DISCUSSION

Characterization of Fe nanowires

AAO template, in which Fe nanowires were electrochemically deposited, was observed by FSEM, as shown in figure 2. It can be seen that the nano-holes in AAO template have been filled obviously. Their diameter is 50 nm, and their density is 1.0×10^6 mm^{-2}.

The magnetism of Fe nanowires was analyzed by SQUID, as shown in figure 3. The coercive force with magnetization perpendicular to the plane of AAO template is 1205 Oe, and that parallel is 401 Oe. It indicates that Fe nanowires arrays have a preferential magnetic orientation along Fe nanowires axis. The magnetization direction is uniaxial magnetic anisotropy because of shape anisotropy of Fe nanowires.

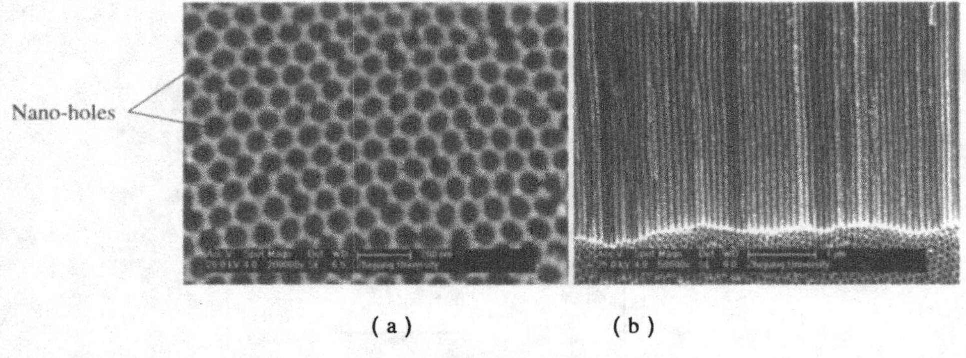

Figure 2. FSEM of AAO template
(a) top view; (b) section view

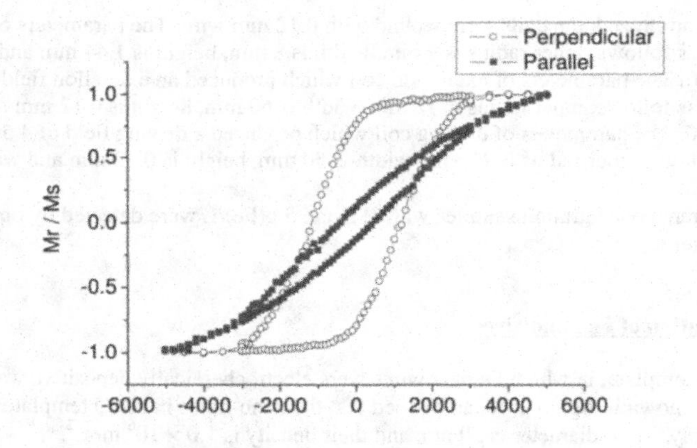

Figure 3. Hysteresis loops of Fe nanowires

Fe nanowires were observed by TEM, as shown in figure 4. The diameter was 50 nm, and the length was 750 nm after ultrasonic. The diameter of the nanowires is equal to the diameter of the nano-holes in AAO template.

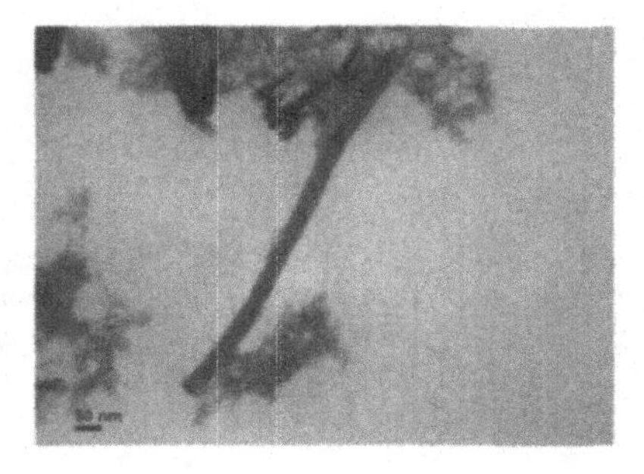

Figure 4. TEM of Fe nanowires

Mechanism of Fe nanowires/chitosan

Through ultrasonic agitation, Fe nanowires are dispersed equably in the mixture of span-80 and liquid paraffin, and the water-in-oil microemulsion system is formed. After Fe nanowires' absorption of chitosan, the glutaraldehyde added in reacts with it, that is to say, carbonyl of glutaraldehyde reacts with amidogen of chitosan to produce schiff bases (- C = N -). The reaction makes the spiral and puckered structure of chitosan molecule more stable and the chain of chitosan molecule more ordinal. This performs well for the covalent conjugation of protein with chitosan.

Experiment of magnetic detecting

Different concentration samples with 1, 2 and 5 g(Fe)/L were detected by our magnetic detection system and the output was depicted, as shown in figure 5. The lowest concentration of samples in this experiment is 1 g(Fe)/L. Through analyzing signal noise ratio, it is determined that the theoretical lower detection limit is 0.2 g(Fe)/L.

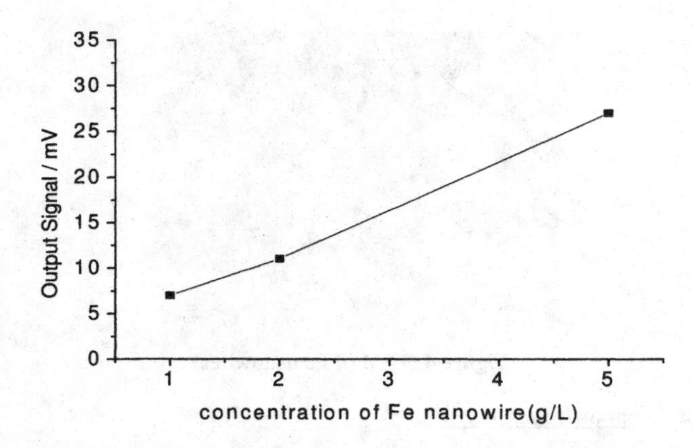

Figure 5. Signal variation of sample with different concentration

The lower detection limit is still higher than the theoretical noise limit given by physical noise limitations, such as the Johnson noise of the coil resistance at room temperature. It is expected that by implementation of all optimization criteria, and especially by excellent balancing, the performance of the magnetic detection system will approach the theoretical noise limit.

CONCLUSIONS

AAO template was prepared by using anodizing voltage step-decreasing method after two-step oxide method. The diameter of the nano-holes in AAO template is 50 nm, and their density is 1.0×10^6 mm^{-2}. Based on AAO template, Fe nanowires arrays were electrochemically deposited. The diameter of Fe nanowires is 50 nm, and the length is 750 nm. Fe nanowires were coated by chitosan. Fe nanowires/chitosan was synthesized by glutaraldehyde as cross-linking reagent. By crosslinking α-human chorionic gonadotropin (α-HCG), biological probes with Fe nanowires/chitosan/antibody were prepared. An easy operating, easy taking and rapid reacting magnetic detecting system was developed after optimizing the geometry parameters of detect coil. Different concentration samples with 1, 2 and 5 g(Fe)/L were detected. The results show that the theoretical sensitivity of system is 0.2 g(Fe)/L and can be improve better.

ACKNOWLEDGMENTS

Research supported by the Key Projects in the National Science & Technology Pillar Program during the eleventh five-year plan period of China (No. 2006BAD30B03)

1. A. Cavalcanti, B. Shirinzadeh and L. C. Kretly. Nanomedicine: Nanotechnology, Biology and Medicine. 4, 127-138 (2008).
2. A. Gomez-Hens, J. M. Fernandez-Romero and M. P. Aguilar-Caballos. TrAC Trends in Analytical Chemistry. 27, 394-406 (2008).
3. W. C. Maki, N. N. Mishra, E. G. Cameron, B. Filanoski, S. K. Rastogi and G. K. Maki. Biosensors and Bioelectronics. 23, 780-787 (2008).
4. M. H. F. Meyer, H.-J. Krause, M. Hartmann, P. Miethe, J. Oster and M. Keusgen. Journal of Magnetism and Magnetic Materials. 311, 259-263 (2007).
5. M. H. F. Meyer, M. Stehr, S. Bhuju, H.-J. Krause, M. Hartmann, P. Miethe, M. Singh and M. Keusgen. Journal of Microbiological Methods. 68, 218-224 (2007).
6. S. Ferretti, S. Paynter, D. A. Russell, K. E. Sapsford and D. J. Richardson. TrAC, Trends Anal. Chem. 19, 530~540 (2000).

Sensing and Detection on Chip – Cells and Particles

Mater. Res. Soc. Symp. Proc. Vol. 1191 © 2009 Materials Research Society 1191-OO11-01

Doped ZnO Colloids for Cancer Detection - Bio-Imaging and Cytotoxicity Study

Linda Y.L. Wu[1,2], G.J. Loh[3], S. Fu[3], A.I.Y. Tok[2], X.T. Zeng[1], L.C. Kwek[3], and F.C.Y. Boey[2]

[1]Singapore Institute of Manufacturing Technology
71 Nanyang Drive, Singapore 638075, e-mail:ylwu@simtech.a-star.edu.sg
[2]School of Materials Science and Engineering
Nanyang Technological University, 50 Nanyang Avenue, Singapore 639798
[3]National Institute of Education, Nanyang Technological University, 1 Nanyang Walk, Singapore 637616

ABSTRACT

We report the synthesis and surface modification of bio-friendly ZnO based colloids, which have been used for cancer cell detection providing significant advantages on quantum confinement effects, high emission brightness in UV to blue-violate range, non-toxicity and a unique dual color imaging feature. The ZnO nanoparticles were single crystal nanoparticles having spherical shape in size of 1-2 or 4-5 nm depending on the surface capping agents. All the colloidal solutions were stable for 30-45 days. The surface capping is a more effective technique in controlling the nanoparticle size, while dopants are effective in modifying the bandgap and optical properties. Unique dual colour images with blue colour in nucleus and turquoise colour in cytoplasm were obtained using either pure ZnO or Co doped ZnO colloids on human osteosarcoma (Mg-63) cells. The dual colour function is the combined effects of quantum confinement and the bio-compatible surface capping groups. The cytotoxicity study proved no cell proliferation by the nanoparticles up to the concentration of 1000 μg/mL, which is the highest concentration reported so far. Since a dosage of only 50-100μM is enough for the *in vivo* detection on rate, these ZnO colloids have high potential for use as the detection media for Lab-on-a-Chip devices.

INTRODUCTION

Semiconductor quantum dots (QDs) offer high quality optical imaging and various particle-cell immobilization advantages over organic markers [1-3]. The major issue with Cd-containing QDs is the potential cytotoxicity, therefore, alternative materials that do not contain cadmium and are more bio-compatible are required [4]. ZnO is a good candidate due to its biocompatibility and versatile optical properties with a wide bandgap (~3.37eV) and an extremely large exciton binding energy (60 meV), which makes the exciton state stable at room temperature and above. Despite the fact that many synthesis methods for obtaining ZnO nanocrystals in aqueous solution at low temperature have been reported [5-7], the required conditions for bio-imaging are not simultaneously satisfied. Common problems of the existing processes for making ZnO nanocrystals are: high temperature or too fast chemical reaction leading to surface defects (Oxygen or Zinc vacancies) resulting in poor optical properties, no suitable surface capping leading to controlled particle size and shape in aqueous solution, toxic ligands used in the process leading to post treatment and potential contamination/toxicity of the final colloidal solution, and no suitable doping leading to emission in visible wavelength range, which is not detectable by common confocal microscopy. Therefore, the major tasks for making ZnO based bio-markers are: (1) to modify the bandgap of ZnO by doping with suitable cations to obtain photoluminescence (PL) emissions in visible wavelength range, (2) to ensure well controlled shape and size of the nanoparticles with particles size in the quantum confinement range (below 10 nm), (3) to provide surface functionalities to ensure the uptake of nanoparticles by the cells, (4) to confirm the photoluminescence emission from the cells after they have absorbed the bio-markers, and (5) to confirm the non-toxicity of the bio-markers. Different dopants (Mg, Ni, Cu, Co, Mn, In, Al, Li, Na, and K) for ZnO have been reported to improve the electrical and ferromagnetic properties [8-12]. For bio-imaging applications, the preferred dopants should have similar atomic radii with Zn, and can reduce the bandgap of ZnO to enhance photoluminescence emission in visible wavelengths. In this article, we report the chemical synthesis and surface modification of bio-friendly ZnO nanocrystals doped with Co, Cu and Ni cations in methonal solution (and changed to water before bio-imaging). Four types of surface capping agents, 1) aminoethyl aminopropyl trimethoxysilane (Z60), 2) aminoethyl aminopropylsilane triol homopolymer water solution (Z61), 3) 3-aminopropyl triehtoxysilane (APTES), and 4) Titania (TiO_2) solution prepared by sol-gel route, were used for the control of particles size and further the bio-conjugation to cells. The nanoparticles were used for *in vitro* bio-imaging of cancer cells and *in vivo* test on rat models. Cytotoxicity was tested on human osteosarcoma cells, and evaluated on rat organ.

EXPERIMENTAL DETAILS

ZnO colloidal solutions were synthesized as described in our earlier publication [13]. In brief, zinc acetate dihydrate, $Zn(Ac)_2 \cdot 2H_2O$ were dissolved in methanol in molar ratio of 0.03:4 and refluxed at 66-67°C for 6-7 hours. Nickel, Cobalt and Copper dopants were added with different molar ratios (x = 0.05, 0.1, 0.15, 0.2) during the synthesis. Capping agent in the calculated amount that is just enough to cover the ZnO particles surface with a monolayer [14] was added. The capping agents include: aminoethyl aminopropyl trimethoxysilane (Z60), aminoethyl aminopropylsilane triol homopolymer water solution (Z61), and TiO_2 solution made from sol-gel route. All the colloidal solutions were exchanged with pure water by distillation before bio-applications. Particle size in solution and size stability were measured by a Malvern Zetasizer Nano ZS, which is equipped with a 4mW 633nm laser. Photoluminescence spectra were measured at room temperature using the colloidal solutions by a single phonon counting spectrofluorometer (Fluorolog-3 Jobin Yvon Horiba).

Bio-imaging and cytoxcicity tests were conducted on human cell lines (osteosarcoma MG-63 cells from ATCC). The cells were cultivated in a glass based round culture dish (diameter 40 mm) in the ATCC complete growth medium containing 10% fetal bovine serum and 1% antibiotics for one day. When the cells occupied more than 50% of the dish area, the culture medium was aspirated and the cells were rinsed by PBS buffer solution. Then 3 μl of ZnO colloidal solution (at concentration of 12.2 μg/μL) was added. Bio-imaging was carried out within four hours by the Nikon C1Si Spectral Imaging Confocal Laser Scanning Microscope System. For cytotoxicity study, Mg-63 cells were seeded in a 96-well flat bottom culture flask with 100 μl per well Dulbecco's modified Eagle's medium (DMEM), and incubated at 37°C for 24 hours. Nanoparticles were diluted in purified water in required concentration, and 10 μl nanoparticle solution was added into each well. The 96-well flask was incubated for another 24 hours. Then 20 μl CellTiter 96 AQ One Solution Cell Proliferation Assay (MTT/MTS assay) was added to each well and incubated for 3 hours. The flask is wrapped in aluminum foil to prevent exposure to light. Absorbance at 490nm was recorded using a 96-well plate reader. Cell viability was calculated by taking the percentage between absorbance of the test sample to the control well without nanoparticle added. *In vivo* tests were performed in rats using two QDs (ZnO-5%Cu-Z60-H2O and ZnO-5%Co- Z60-H2O). 6-8 weeks old male rats were weighed, labeled and injected intraperitoneally with the QDs in concentration of 50μM. After 1 or 24 hours, the rats were given 3ml anesthesia consisting of ketamine/diazepam, and fluorescence at the rats' external body was analysed using an Ocean Optics USB 2000 fiber optic spectrofluorometer. The rats were then dissected to measure the fluorescence spectrum *in vivo* from four organs i.e. liver, kidney, lung and spleen. Five data points were taken from each organ and an average was computed. Tissue specimens from the liver, kidney, lung and spleen were taken for fixation and embedding in paraffin. For fixation, 10% buffered formalin (i.e. 4% solution of formaldehyde) was used. An automatic tissue processor, Leica TP1020, was used to process the fixed tissue specimens. The specimens were then sectioned to 6μ thickness using Leica RM2135 rotary microtome and placed onto microscope slides for taking the fluorescence images using the Leica True Confocal Scanner (TCS) SP2 Confocal Laser Scanning Microscopy (CLSM).

RESULTS AND DISCUSSION

Particle size and colloidal stability

The particle size and colloidal stability were evaluated by Zetasizer and reported earlier [14]. The colloids were stable for 30 to 45 days. Regardless of the dopants used, the most stable colloids were capped by Z60, which were stable for more than 45 days. This indicates that the double amino- groups contained in the Z60 have better capping ability than the single amino-group contained in APTES. The homopolymer water solution (Z61) made the particles stable for 30 days, and the water based TiO_2 sol-gel solution led to slightly larger size from beginning and less stability.

Photoluminescence spectra

The bandgap modifications by different dopants were analysed and reported earlier. In order to correlate the PL emission with the bio-imaging colors, Fig. 1 shows the PL spectra of the 5%Co-ZnO without capping (H_2O only) and with Z60, Z61 and TiO_2 capping agents. It is seen that uncapped 5%Co-ZnO emits at about 390 nm, TiO_2 capped 5%Co-ZnO emits at 380 nm, while Z61 and Z60 capped 5%Co-ZnO emit at 410 and 420 nm respectively. The influences of surface capping and doping of ZnO on the PL properties should be discussed from the mechanisms of photoexcitation and emission of semiconductor materials. Upon photoexcitation of a semiconductor particle, an electron-hole pair is created. This pair can exist as a Wannier exciton and when it recombines, emits a photon with an energy close to the bandgap of the material. This is referred to as exciton emission. The electrons and holes can also be trapped somewhere in the particle especially at dopants substituted sites. The energetic position of a shallowly trapped charge carrier is related to the band structure, which is a function of particle size (so called quantum confinement effect). The energetic position of a deeply trapped charge carrier is independent of particle size and is determined by the chemical nature of the trap and by the local structure surrounding the trapped charge carrier. Emission of trapped charge carrier can occur via a radiative process (called trap emission) or a non-radiative process (called non-radiative recombination). Exciton emission, trap emission and non-radiative recombination are the three competing processes that determine the final PL spectrum [15].

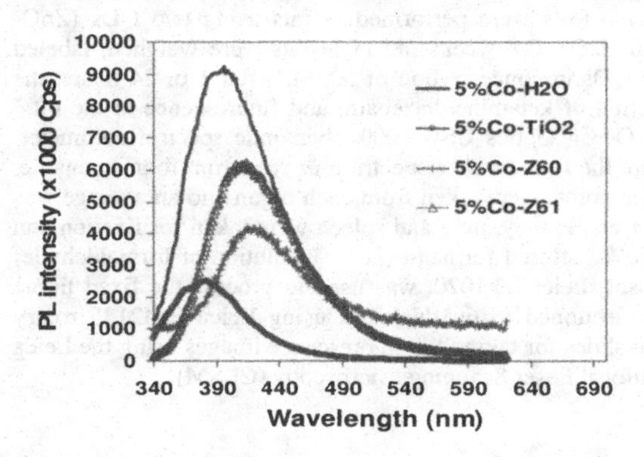

Fig. 1 PL spectra of 5%Co doped ZnO, uncapped and capped with Z60, Z61 and TiO_2 respectively in water solutions, excited at 325 nm.

Bio-imaging on human cells

Figure 2 shows the bio-images of MG-63 cells labeled by 5%Co-ZnO-Z60 (a) and uncapped 5%Co-ZnO (b) respectively. It is seen that a dual color image is obtained using 5%Co-ZnO-Z60 with a blue color in the nucleus and a turquoise color in the cytoplasm. Uncapped 5%Co-ZnO labeled cells also show dual color emission but the colors are blue shifted. This could be related to the PL spectra in Fig.1. The PL emission of uncapped 5%Co-ZnO is at 400 nm (blue-violate color), while Z60 capped 5%Co-ZnO emits at 420 nm (blue turquoise color). The colors of the images are very well matched with the PL emission peaks. The unique dual color imaging feature is believed due to the small particle size (1-2 nm), the double amino surface functional groups, which have high affinity to the cells, and the high PL emission intensity. This feature eliminates the complexity of two-step labeling using two different markers and the need of two excitation sources in the case of fluorescent dyes. Cu and Ni doped ZnO labeled cells were not detected dual color emission, which may be due to the less absorbed nanoparticles by the cells.

Cytotoxicity test results

The cell viability after adding different nanoparticle concentrations (400 to 1000 µg/mL) are plotted in Fig. 3. Only 5%Cu and 5%Co doped ZnO results are plotted as these results were more consistent. It is seen from the figure that Aptes capped nanoparticles are slightly toxic to cells. TiO_2 capped Cu doped ZnO are also slightly toxic, and all the Z60 and Z61 capped ZnO are safe (viability higher than 0.5) up to 1000 µg/mL concentration. Concentrations beyond 1000 µg/mL were not tested because this concentration is already much higher than the required concentration for both *in vivo* and *in vitro* bio-imaging applications. Reported injection concentrations for mouse *in vivo* tests are in the range of 10 to 250 µg/mL [16-18], and maximum concentration for *in vitro* tests on live cells is 100 μg/mL [19,20]. Cytotoxicity of Cd-containing QDs was already found in 62.5 μg/mL concentration. The toxic threshold for cells is consistently reported to be in the tens of p.p.m. range (tens of μg/mL) [21]. Therefore, our concentration of 1000 μg/mL is the highest concentration reported so far.

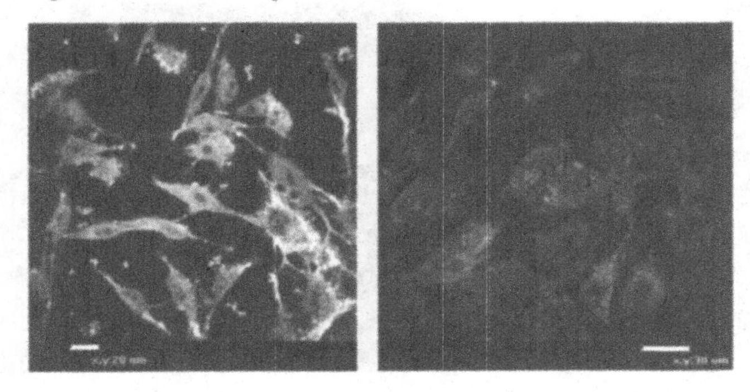

Fig. 2 Bio-images of MG-63 cells labeled by 5%Co-ZnO-Z60 (a) and uncapped 5%Co-ZnO (b).

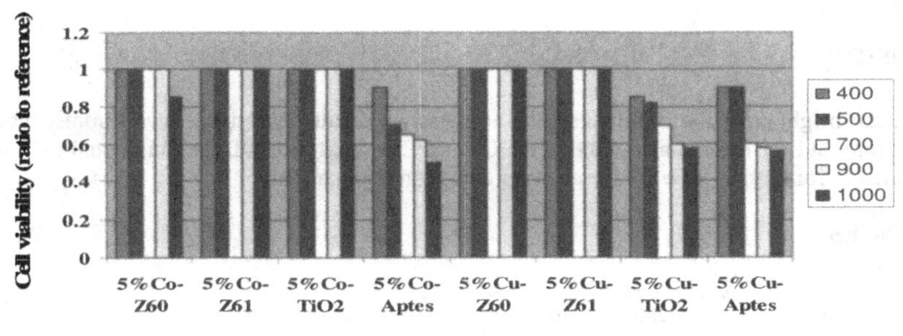

Fig. 3 Cell viability results tested by MTT method using 5%Co doped ZnO and 5%Cu doped ZnO.
Legends: concentration of nanocrystals added in unit of μg/mL. Standard deviations of the values are about 10%.

In vivo test in rat

The results shown in Fig.4(a) is the PL emission curve from the lung tissue of rat which was injected with ZnO-5%Co-Z60-H_2O (50μM) and culled after 1 hour (C1) and 24 hours (C2) respectively compared to the control rat without QDs added (A1). The higher fluorescence intensity of C1 curve indicates a high absorption of QDs by the lung in just one hour. C2 curve indicates a drop of intensity after 24 hours due to absorption of QDs into deep tissue in the lung hence not detectable from the tissue surface. The corresponding confocal image of the lung is shown in Fig. 4(b). Since no negative effect was observed on rat after injection of QDs, these results confirmed the suitability of the QDs for *in vivo* test on live animals. The nanocrystals can be easily attached with proteins or antibodies for specific cancer detection using a Lab-on-a-Chip device.

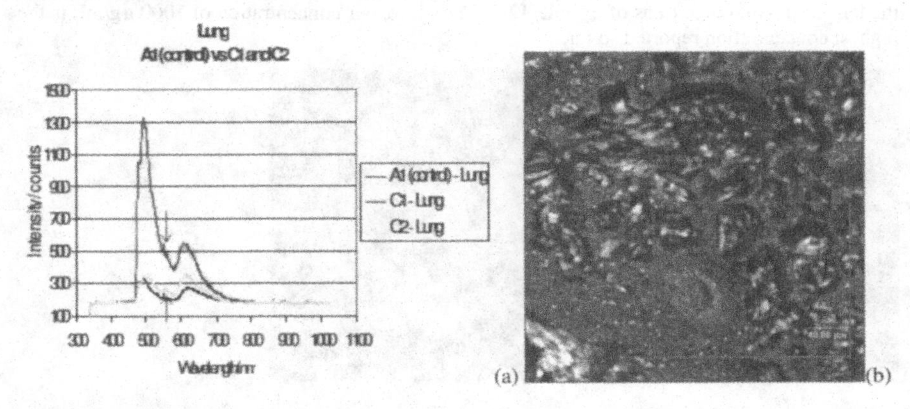

Fig. 4 Fluorescence spectrum (a) and image (b) from lung of rat injected with ZnO-5%Co-Z60-H2O (50μM) QDs. C1 was taken after 1 hour of injection and C2 was taken after 24 hours of injection. (b) was taken after 24 hours.

CONCLUSION

The dual color bright emission bio-images, the non-toxicity results, and the *in vivo* fluorescence emissions from rat lung have confirmed that the Co and Cu doped ZnO QDs are high quality bio-markers suitable for both *in vitro* and *in vivo* bio-detection applications.

REFERENCES

[1] Politz, J.C. *Trends in Cell Biology* **9**(7), 284-287 (1999).
[2] Bruchez Jr, M., Moronne, M. *Science* **281**(5385), 2013 (1998).
[3] Alivisatos, A.P. *Science* **271,** 933-937 (1996).
[4] Lin, C.J., Liedl, T., Sperling, R.A., Fernandez M.T., Pereiro, R., Medel, A., *et al. J. Mater. Chem.*, **17,** 1343-1346 (2007).
[5] Ristic, M., Music, S., Ivanda, M., Popovic, S. *Journal of Alloys and Compounds* **397**(1-2), L1-L4 (2005).
[6] Bauermann, L.P., Bill, J., and Aldinger, F. *J. Phys. Chem. B* **110,** 5182-5185 (2006).
[7] Soares, J.W., Steeves, D.M., Ziegler, D., DeCristofano, B.S. *Proceedings of SPIE* , **6370,** 637011 (2006).
[8] Ning, G.H., Zhao, X., Li, J. *Optical Materials* **27,** 1-5 (2004).
[9] Gonzalez, A.E.J. *J. Solid State Chem.* **128,** 176-180 (1997).
[10] Chen, Y.W., Liu, Y.C., Lu, S.X., Xu, C.S., Shao, C.L., Wang, *et al. J. Chem. Phys.* **123,** 134701 (2005).
[11] Cong, C.J., Liao, L., Li, J.C., Fan, L.X., Zhang, K.L. *Nanotechnology* **16,** 981-984 (2005).
[12] Liao, Y., Huang, T., Lin, M., Yu, K., Hsu, H.C., Lee, T., *et al. J. Magnetism Magn. Mater.* **310,** e818-e820 (2007).
[13] Wu, Y.L., Fu, S., Tok, A.I.Y., Zeng, X.T., Lim, C.S., Kwek, L.C., Boey, F.C.Y. *Nanotechnology* **19,** 345605 (2008).
[14] Wu, Y.L., Lim, C.S., Fu, S., Tok, A.I.Y., Lau, H.M., Boey, F.Y.C., Zeng, X.T. *Nanotechnology* **18,** 215604-215612 (2007).
[15] Dijken A.V., Meulenkamp E.A., Vanmaekelbergh D., Meijerink A. *J. Phys. Chem. B* **104,** 1715-1723 (2000).
[16] Derfus A.M., Chan, W. C. W., & Bhatia S.N. *Nano Lett.* **4,** 11-18 (2004).
[17] Gao X.H., Cui Y., Levenson R.M., Chung L.W.K., & Nie S. *Nat. Biotechnol.* **22**(8), 969-976 (2004).
[18] Ballou B., Lagerholm B.C., & Ernst L.A. *Bioconjugate Chem.*, **15,** 79-86 (2004).
[19] Selvan S.T., Tan T., & Ying J.Y. Robust, *Adv. Mater* **17,** 1620-1625 (2005).
[20] Hoshino A., Fujioka K., Oku T., Suga M., Sasaki Y.F., Ohta T., *et al. Nano Lett.* **4**(11), 2163-2169 (2004).
[21] Clarke S.J., Hollmann C.A., Zhang Z., Suffern D., Bradforth S.E., Dimitrijevic N.M., *et al. Nat. Mater.* **5,** 409-417 (2006).

Mater. Res. Soc. Symp. Proc. Vol. 1191 © 2009 Materials Research Society 1191-OO11-05

Screen-Grid Field Effect Transistor for Sensing Bio-Molecules

Kwee G. Eng[1], Kristel Fobelets[1] and Jesus E. Velazquez-Perez[2]

[1]Department of Electrical and Electronic Engineering, Imperial College London, Exhibition Road, SW7 2AZ London, UK
[2]Departmento de Fisíca Aplicada, Universidad de Salamanca, Edificio Trilinüe, Pza de la Merced s/n, E-37008 Salamanca, Spain

ABSTRACT

A novel field effect transistor, based on the Screen Grid Field Effect Transistor concept, is proposed with an integrated Coulter Counter pore for amplification of the sensing signal. 3D TCAD (Technology Computer Aided Design) simulations are performed on the use of the Coulter Counter Field Effect Transistor (CC-FET) to detect the Influenza A virus. The gate of the transistor is the pore through which the bio-particles pass. This passage causes a change in the electrostatic conditions of the gate and thus changes the source-drain current, similar to ISFET (Ion Sensitive FET) operation. The structure of the CC-FET is optimized for bio-sensing and multi-particle passage through the gate hole is simulated. TCAD results show that the CC-FET is capable of multi-particle and particle size detection.

INTRODUCTION

Improved healthcare is a key objective in modern society. Medical diagnostic systems play an important role to establish quickly and accurately the underlying causes of an illness. A wide range of biosensors [1] have been and are being developed including the ISFET (Ion Sensitive Field Effect Transistor) [2] and Coulter Counter [3], both relevant to this work. Biosensors determine the number and type of extremely small bio-particles, e.g. red blood cells, viruses, etc. Research on the Coulter counter (CC) has seen a revival recently because of the improved control of the pore size and homogeneity via synthetic pores fabricated with microelectronics technology [4]. CCs consist of a membrane with a small pore through which the bio-particles distributed in an ionic liquid pass. Each time this happens, a sudden change in current or voltage is measured. These signals are small in amplitude and noisy. Signal processing techniques can be employed for optimizing the signal. In this work we propose to integrate the CC with a Field Effect Transistor (FET) in order to improve the signal by combining the CC and FET response. We investigate, using TCAD (Technology Computer Aided Design) [5] the operation of this integrated device. This combined CC-FET (Coulter Counter - Field Effect Transistor) structure will also open the possibility of densely packed integrated sensing arrays.

In order to understand the functioning of the CC-FET, we introduce the principle of the Screen-Grid Field Effect Transistor (SGrFET). The SGrFET [6] has a completely novel gating structure that is composed of metal/oxide filled vacuoles in the Si channel of a FET on SOI (silicon-on-insulator). The vacuoles are from the top of the SOI to the buried oxide layer. Their inner surface is covered with a gate oxide and then further filled with metal. The principle of operation of the SGrFET is analogous to the MESFET (Metal Semiconductor FET) [7] and is based on the degree of depletion between the gate cylinders. Modeling the performance of the SGrFET using TCAD has shown promising results for different applications [6]. Due to the specific vacuole-based gating structure through the channel region of the SGrFET it can be used for bio-sensing applications, provided that the gate vacuoles are opened through the substrate. The bio-particles travel through the gate vacuole(s) similar to the Coulter counter but as the pore is an active FET an additional current can be measured [8]. This approach gives two electrical signals per device (one from the CC and one from the FET), making sensing more robust. Fig. 1 gives an illustration of the SGrFET when used as a CC-FET. A bio-particle in the gate hole changes the electrostatic condition in the channel of the SGrFET similar to a bioFET [9].

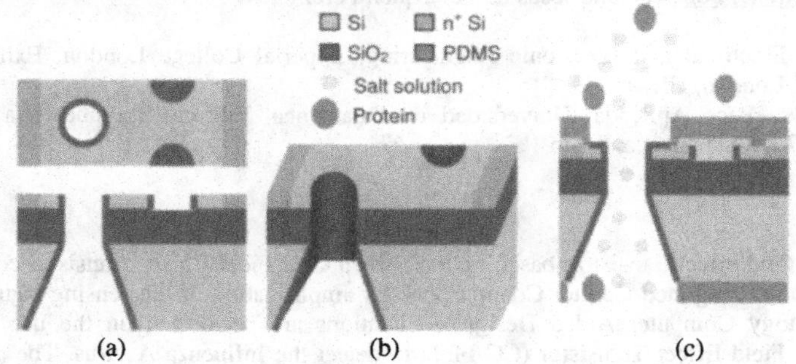

Figure 1: (a) 2D top and cross section of a CC-FET with the left hole processed as a pore and the right holes acting as control gates. (b) 3D cross section. (c) CC-FET in an electrolyte with proteins.

3D TCAD – Taurus from Synopsis [10] – is used simulate the operation of the CC-FET. First the structure of the SGrFET is adapted and optimized for use as a CC-FET. Then the operation of CC-FET is simulated for particles flowing through the gate vacuole.

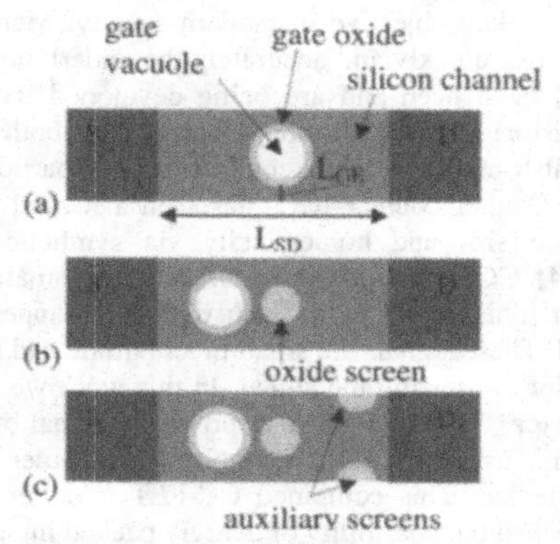

Figure 2: Schematic top view of the CC-FET. S, D: ohmic contact regions at the edge of a Si channel. Circles are vacuoles filled or partially (white) filled with SiO_2. L_{SD} is the source to drain distance and L_{GE} is the channel width. Designs from (a) to (c) give improved FET sensitivity but also increased fabrication complexity.

RE-DESIGN OF THE SGRFET INTO A CC-FET

The classical SGrFET consists of a cell with 4 gate cylinders. The Coulter counter principle however is based on one pore, thus the SGrFET needs to be re-designed with one gate pore while maintaining excellent field effect transistor performance. The schematic top views of a single pore and improved CC-FET structures are given in fig. 2. It was found that the characteristics of the SGrFET with only one gate pore, see fig. 2(a), were much degraded compared to the full double gate row SGrFET presented in [6]. The excellent operation of the SGrFET is based on the electrostatic source potential shielding by the 2nd row of gates and the multi gate action, both of which are removed in fig. 2(a). In order to improve the CC-FET performance, additional shielding is needed in the channel region. Two improvements were investigated; first an oxide screen was introduced behind the pore to prevent the drain from controlling the charges in the channel (fig. 2(b)). In the second this shielding was increased via the introduction of two auxiliary oxide screens close to the drain contact (fig. 2(c)).

Increasing the screening as in fig. 2 (b,c) increases the transconductance efficiency – thus providing increased sensitivity to the electrostatic influence of the bio-particle. Fig. 2(c) shows the best sensitivity. However, increasing the number of oxide screens increases the fabrication complexity. Metal shielding could give better performance but will further increase the fabrication complexity. A further improvement to particle sensing can be made by introducing an electrostatic shield above and below the channel region as shown in fig. 3. A grounded electrode or highly doped silicon layer is used as electrostatic shield, leaving only the walls of the pore exposed to the electric field. To prevent the electrode shield from forming an ohmic contact with the silicon channel and short circuiting the source and the drain, a thin layer of oxide is defined between the electrode and the channel. This shielding allows a sequence of crossing particles to be sensed independently.

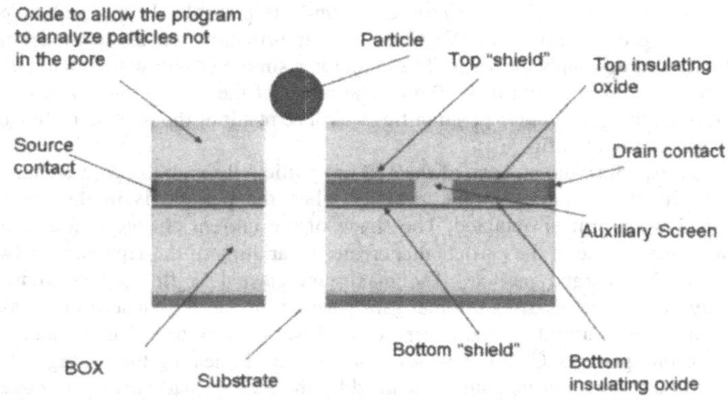

Figure 3: Cross section, through a pore, of the critical modifications carried out to the CC-FET for improved shielding. The silicon substrate is only partially shown. The particle is a model for the influenza A virus of the same size and charge.

PARTICLE SENSING USING THE CC-FET

Device simulators such as the one used in this work have become standard in the semiconductor manufacturing world. A simulation approach minimizes the cost of fabricating novel device structures. These TCAD programs are oriented towards the electronics industry and do not provide a function to simulate movement. In our approach we simulate movement of the particle through the gate pore by calculating the response of the CC-FET to different positions of the particle relative to the gate pore. Therefore a separate structure input file for each position of the "moving" particle is created and the response calculated. The results from each position are then combined and plotted on a graph as a function of the various particle positions. In fig. 4 the sequence of TCAD generated pictures is given, showing the particle's simulated travel through the gate pore.

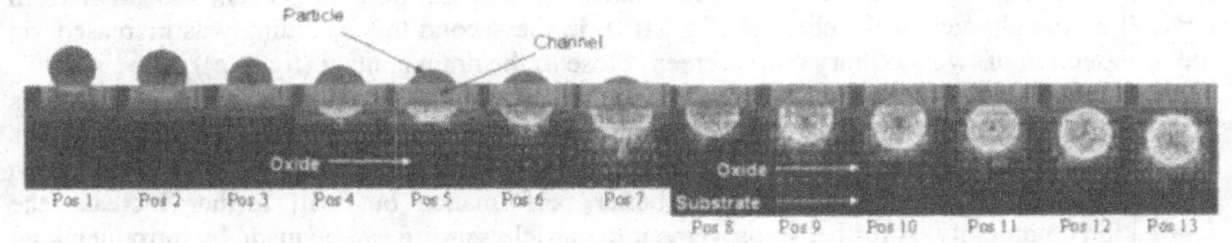

Figure 4: Sequence of pictures depicting the simulated passage of a particle through the gate vacuole. Pictures are created by the TCAD program [10].

A bio-particle, useful for detection and simulation, is the influenza A virus. The influenza A virus is a spherical nano-particle with a 80-120 nm diameter. Different molecules form the basic elements for these complex bio-particles which exhibit electrical properties akin to charged particles of 2 electron units. Depending on the pH level of the liquid the virus is mixed in, an optimum zeta potential [11] of -80mV to -100mV is possible. In a liquid of pH≈7 the zeta potential is approximately -100mV. Using this information we use a model particle in the simulations that represents the virus. The model is a simple sphere with a diameter of 120 nm at a uniform electrical potential of -100mV. The result of the simulation, when only one particle passes through the gate hole, is given in fig. 5(a). The result of the passing of five closely packed particles can be found in fig. 5(b).

From fig. 5(a), the passage of the 120 nm particle through the pore creates an observable change in the drain-source current, peaking when the particle is in the channel. A nearly symmetrical waveform is obtained. The shape of the current change is a consequence of the spherical characteristic of the particle that creates a variation of the separation between gate pore surface and the charged particle. The maximum current is flowing when the particle sits completely in the pore ensuring optimal gating action. In fig. 5 (b) a series of peaks and troughs in the drain-source current can be observed. Each set of peak-trough combinations indicate the presence of one particle. Clearly five sets can be seen, indicating the passage of five particles. We notice that the current magnitude induced by the 120 nm particles is the same regardless of the number of particles passing through. This is a significant development as the size of the particle can be obtained from the amplitude of the current and no additional signal processing is required. In fig. 6 the passage of three particles with different diameter: 120 nm, 100 nm and 80 nm, is shown. To simulate the random speed at which particles move through the pore, the relative distance of each particle was varied, thus the peaks are not equidistant.

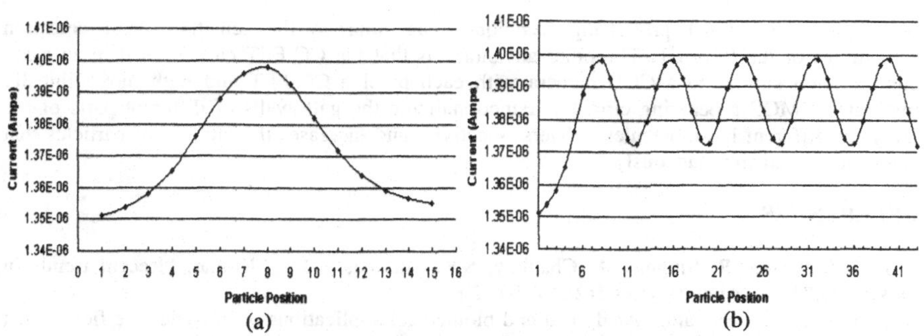

(a) (b)

Figure 5: (a) The current magnitude waveform when a single particle transits the pore. (b) The current magnitude waveform when five particles of the same type pass the pore.

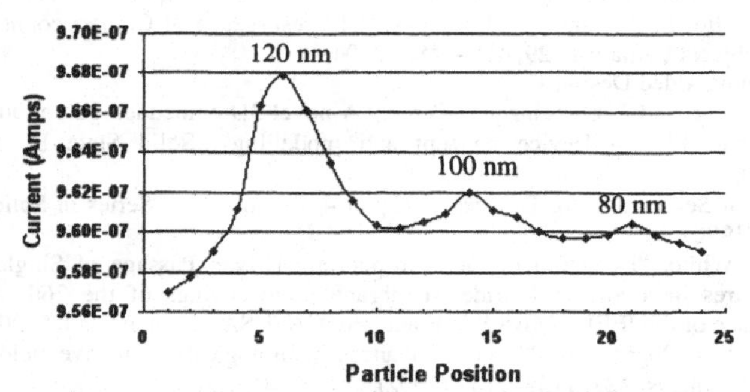

Figure 6: Response of the current to a continuous stream of particles of different sizes (120, 100 and 80 nm) passing through the pore at different distances from each other.

Three distinct peaks are observed, each with a different current magnitude that corresponds exactly to their single particle passage peak current. Hence, we can conclude that each particle of a different size induces a distinct current magnitude, regardless of the number of particles going through or how close they are to each other. The highest current changes are obtained by the particles with sizes most closely matched to the pore diameter. This attribute can be used by the processor to determine the size of particles flowing through the pore.

DISCUSSION AND CONCLUSION

Simulations have shown that the CC-FET is able to detect and count the passage of particles similar to the influenza A virus through the gate vacuole. The size of the particle can be directly measured via the magnitude of the current peak. The flow of the particles can be increased via the use of a micropump as the cut-off frequency of the active CC-FET is tens of MHz. Two additional advantages of the use of the CC-FET can be proposed. The first is the combined detection of the traditional CC signal with the CC-FET signal. The combination of the two signals via signal processing techniques can improve the sensitivity and detection capabilities of the CC-FET. The other advantage is that the CC-FET can be used in an array configuration similar to a CCD camera with each pixel a CC-FET and with monolithically integrated CMOS processing circuitry. Functionalizing the gate walls of different parts of the array for different bio-molecules delivers selectivity and increases the number of particles that can be measured simultaneously.

REFERENCES

1. B.D. Malhotra, R. Singhal, A. Chaubey, S.K. Sharma, and A. Kumar, "Recent trends in biosensors," Curr. Appl. Physics **5**(2), 92–97(2005).
2. P. Bergveld, A. Sibbald, "Analytical and biomedical applications of ion-selective field-effect transistors", Elsevier (1988).
3. R.W. DeBlois, and C.P. Bean, "Counting and sizing of submicron particles by resistive pulse technique". Rev. Sci. Instrum. **41**, 909–916 (1970).
4. R. R. Henriquez, T. Ito, L. Sun and R. M. Crooks, "The resurgence of Coulter counting for analyzing nanoscale objects", Analyst 129**,** 478 – 482 (2004).
5. Technology Computer Aided Design
6. K. Fobelets, P.W. Ding, and J. E. Velazquez-Perez, "A novel 3D embedded gate field effect transistor – Screen-grid FET – Device concept and modelling", Solid State Electronics **51**(5), 749-756 (2007).
7. M. Shur, "Physics of Semiconductor Devices", chapter 4, Prentice Hall Series in Solid State Physical Electronics (1990).
8. S. Pandey, M. H. White, "Detection of Dielectrophoretic Driven Passage of Single Cells through Micro-Apertures in a Silicon Nitride Membrane", Proceedings of the 26th Annual International Conference of the IEEE EMBS San Francisco, CA, USA • September 1-5, 2004.
9. M. J. Schöning and A. Poghossian, "Recent advances in biologically sensitive field-effect transistors (BioFETs)", Analyst **127**, 1137–1151 (2002).
10. http://www.synopsys.com/
11. "The Zeta (ξ) potential is the electrostatic potential that exists at the shear plane of a particle, which is related to both surface charge and the local environment of the particle" http://www.fen.bilkent.edu.tr/~ozbay/Papers/166-08-biomicrodevice.pdf

Mater. Res. Soc. Symp. Proc. Vol. 1191 © 2009 Materials Research Society 1191-OO11-06

Resonant microcavities coupled to a photonic crystal waveguide for multi-channel biodetection

Elisa Guillermain and Philippe M. Fauchet
Electrical and Computer Engineering Department, University of Rochester,
Rochester, NY 14627 USA

ABSTRACT

Miniaturized and highly sensitive bio-sensors are attractive in various applications, such as medicine or food safety. Photonic crystal (PhC) microcavities present multiple advantages for rapid and accurate label-free optical detection. But their principle of operation (i.e. observation of a peak in transmission) makes their integration in serial arrays difficult. We present in this paper a multi-channel sensor consisting of several resonant PhC microcavities coupled to the same waveguide. The transmission spectrum shows as many dips as there are cavities, and each of the microcavities can act as an independent sensor. Preliminary results show the fabrication and characterization of a double-channel structure with small defects used as a solvent sensor.

INTRODUCTION

Many optical sensors rely on the modification of their optical properties when their refractive index changes due to the trapping of the target in the detection area. They allow a label-free detection [1], relying on the red-shift of their optical resonance wavelength. Such bio-detectors include surface plasmon resonance (SPR) [2], Bragg surface wave [3], interferometers [4], waveguides [5], disk resonators [6], photonic crystals (PhC) [7,8,9,10,11,12]. Photonic crystals (PhC), of interest for bio-sensing because of their high sensitivity in a small sensing area (less than 10 μm^2), are studied in this paper.

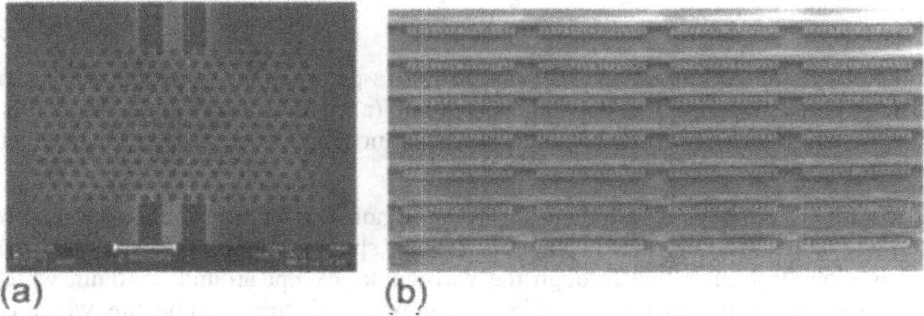

Figure 1. SEM views of (a) a 2D photonic crystal cavity suitable for the sensing of proteins [9]. (b) 1D photonic crystal cavities in a serial configuration allowing multiple detection, proposed by Mandal and Erickson [11].

Fig. 1 shows two types of PhC microcavities for bio-detection applications. The structure showed in Fig. 1(a) is a PhC with a microcavity (smaller hole), whose resonance is in the infrared. At this resonnance wavelength, the electric field of the optical mode is mostly confined in the defect hole, which reduces the detection area to this region. This device has been shown to be suitable for the detection of bovine serum albumin (BSA) attached to a glutaraldehyde functionalization layer. A redshift of 2 nm has been observed after capture of the targeted BSA molecules forming a layer with an effective thickness of 1.5 nm [9]. Such structures have also been shown to be able to detect bigger particles, for example viruses [10].

Fig. 1(b) shows a SEM top view of a waveguide to which several 1D PhC micro-cavities are coupled [11]. This kind of device is suitable for multi-channel detection (simultaneous multiple detections), allowing for example to target multiple molecules, or to avoid false-positive detection. The input light is globally transmitted through the waveguide, except for the wavelengths corresponding to any of the cavity resonant modes. Because each microcavity is slightly different from the others, their resonant wavelengths are different. The transmission spectrum shows as many dips as there are cavities [11], and each cavity can act as an independent sensor.

The PhC microcavities shown in Fig. 1(a) do not allow a serial array configuration. If two microcavities with different resonant wavelengths are placed in series, then the wavelength peak transmitted thought the first cavity will fall inside the photonic bandgap of the second one. This leads to negligible light transmission and detection is impossible. Multi-channel devices detection principle relies on the shift of a narrow transmission dip, not a peak.

Here we propose a PhC structure which maintains all the PhC advantages (i.e. high-sensitivity, miniaturization, possibility of single particle detection [10,12]) and allows for integration in a serial array. Creating a PhC waveguide by removing a line of holes in the PhC allows the light to be globally transmitted. Coupling several 2D PhC cavities to this waveguide would lead to several transmission dips in the optical spectrum, corresponding to the resonance wavelength of each microcavity.

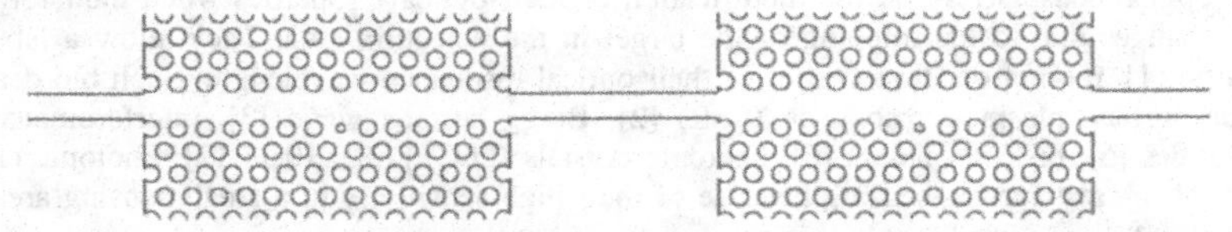

Figure 2. Schematic top view of the proposed structure implemented in a double-channel configuration.

SIMULATIONS

2D simulations were performed in TE polarization (s-polarization) with the FDTD freeware Meep. The structures consist of a W1 PhC waveguide (a missing line of holes in the PhC) [13], where one of the holes near the waveguide has been modified to create a resonant microcavity (as shown in Fig. 2).

The PhCs studied here have a periodicity of 400 nm, holes of 120 nm radius, and a defect hole with a 60 nm radius. The calculated TE transmission spectrum (Fig. 3(a), straight line) shows that light is globally transmitted through the waveguide, except around 1510 nm which is the resonant wavelength of the microcavity. Fig. 3(b) shows its electric field profile, which is mostly confined in the defect hole.

Calculations have also been performed to show the sensitivity of this structure when a bio-material with a refractive index of 1.34 is attached to the walls of the holes. Fig. 3(a) shows that grafting bio-molecules in the structure leads to a redshift of the dip. Fig. 3(c) shows the evolution of the shift with the thickness of bio-molecules grafted. A wavelength redshift of 4 nm is expected for the grafting of a bio-layer with an effective thickness of 5 nm.

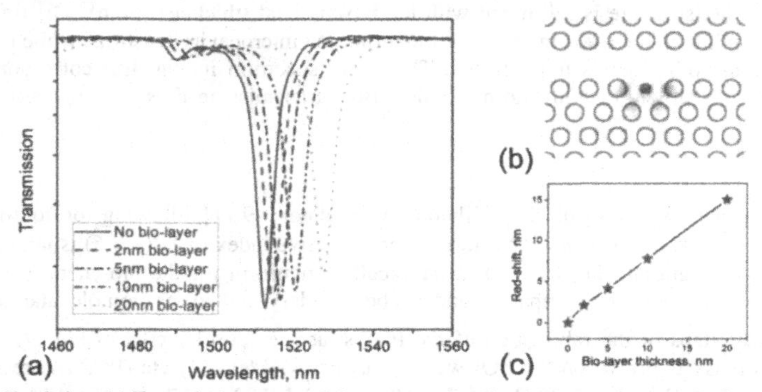

Figure 3. Simulations (a) Transmission spectra of the structure described in the text: in air (solid line) and coated with different thickness of bio-material -refractive index 1.34- (dashed and dotted lines). (b) Electric field resonant mode profile of the structure in air. (c) Evolution of the dip wavelength with bio-material thickness.

By letting most of the light pass through the waveguide, this kind of sensor allows for multi-channel integration. A double-channel device consisting of two different microcavities coupled to the same PhC waveguide, as shown in the Fig. 4(b), has been simulated. The two PhC have periodicities of 400 and 408 nm, holes radii of 120 and 126 nm, and their defect holes are 60 and 63 nm radii respectively. The periodicity of each crystal is slightly different, allowing one to observe two different resonant wavelengths and to keep the same quality factor for both the modes. The upper spectrum in Fig. 4(a) shows the two transmission dips for the structure in air.

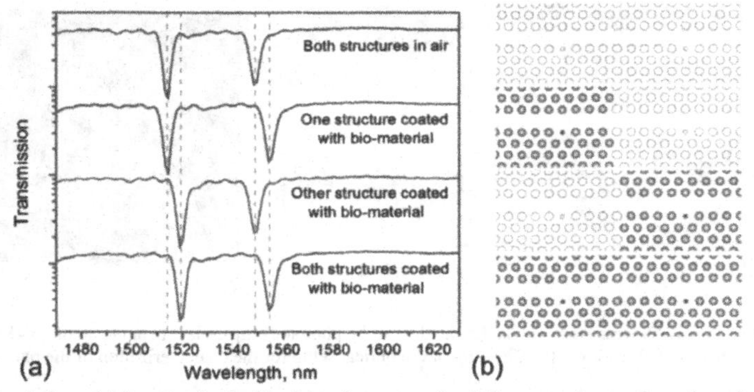

Figure 4. Simulations. (a) Spectra of a double-channel structure, for different sensing configuration when grafting a 5nm bio-layer on the PhC walls. (b) Schematic of the sensing configuration studied.

Simulations have also been performed to show that the proposed device can detect individually the attachment of bio-molecules in each of the defect areas. Fig. 4(a) shows the transmission spectra calculated for different sensing conditions. The upper spectrum concerns the structure in air. The two middle spectra correspond to the double-channel structure transmission if only one side of the PhC structure is infiltrated with the bio-material (thickness 5nm). The transmission dip corresponding to the resonant mode of the infiltrated microcavity redshifts, while the spectral position of the other dip is not affected. The lower spectrum in Fig. 4(a) corresponds to the coating of both structures with bio-molecules. Both transmission dips then redshift, by about 7 nm each.

METHODS

The PhC devices are based on SOI (Silicon On Insulator) [9,11], allowing for confinement of light in the vertical out-of-plane direction, as the refractive index of Si (~ 3.5) is larger than that of SiO_2 (~ 1.5) and air (1). SOI is also an excellent platform for the bio-chemistry needs, as silicon dioxide (glass) is biocompatible and has been widely studied for biomolecular bonding.

The different steps of the fabrication of the PhC structures are depicted in Fig. 5(a). First, an oxide hard mask is grown on the SOI wafer. Polymethyl Methacrylate (PMMA) resist is then spinned on the substrate at 3000 rpm for 60 sec and baked at 170 °C for 15 minutes. The photonic structure pattern is then written on the resist with an e-beam writing system (JEOL JBX-9300FS), using a 2 nA beam and a 1500 C/cm² dose. PMMA is then developed in a 1:3 MIBK:IPA solution, and the holes of the PhC structures are open.

The etching steps, performed by Reactive Ion Etching (RIE), allow the transfer of this pattern to the lower layers. The oxide hard mask is first etched through the resist mask, with an Ar assisted CHF3 plasma (5 min at 200 W). The resist is then stripped by oxygen plasma, and the top oxide layer acts as a hard mask for the silicon layer, which is etched with a Cl2, BCl3 and H2 plasma. This patterned silicon layer is the photonic crystal to be characterized after dicing the chip and polishing its input and output edges.

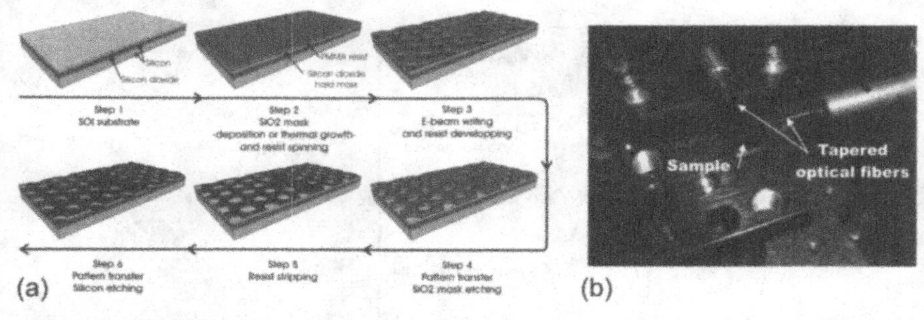

Figure 5. Methods (a) Process steps for the fabrication of the photonic chips, via e-beam technology (PMMA being the resist used) and RIE etching. (b) Close-up of the optical setup for the characterization of the photonic chip.

The optical testing setup shown in Fig. 5(b) is used to measure the transmission spectra of the fabricated devices. The source is a tunable laser (HP 8168F) with a frequency range of 1440 nm to 1590 nm. The light is ensured to be TE polarized (s-polarization) via a polarizer controller. The polarized light exits the optical fiber via a taper, and is coupled into the tapered waveguide of the photonic chip. The light transmitted though the PhC structures is then decoupled from the chip and detected with an InGaAs detector.

PRELIMINARY RESULTS

The devices fabricated and optically tested have a periodicity of 370 and 380 nm with hole radii of 98 and 100 nm respectively. The two microcavities consist of a 60 nm radius hole and a 61 nm radius hole. They are separated by a long waveguide (80 μm) in order to further allow the implementation of each of them with two different microfluidic channels. Fig. 6 shows the measured transmission spectra obtained for three different structures, respectively, from top to bottom, (1) a structure with a 370 nm period, (2) a structure with a 380 nm period, and (3) the double-channel device, with both structures in a serial array (i.e. in the same waveguide).

The solid spectra (1) and (2) -corresponding to single cavities- show that the light is globally transmitted through the structure, except at the cavity resonance wavelength (1516 nm and 1542 nm). The measured quality factors are 400 and 510.

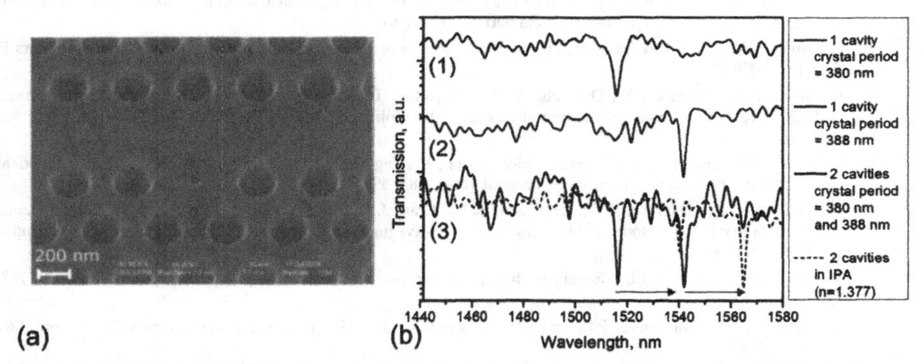

Figure 6. Experimental (a) SEM top view of a fabricated microcavity along the PhC waveguide. (b)Transmission spectra of (1) a structure with a 370 nm period, (2) a structure with a 380 nm period, and (c) the double-channel device, with both structures in a serial array (i.e. in the same waveguide). The solid spectra correspond to the transmission of the structure in air, and the dotted ones to the transmission in IPA (n=1.377).

The spectrum (3) corresponds to the optical transmission of the double-channel device. As anticipated, we observed two dips. The resonant wavelengths for the single cavity structures and for the one on the double-channel device are the same. In the presence of IPA (n=1.377) in the holes, a redshift of the resonance wavelengths of 24 nm and 23 nm is observed, showing a sensitivity of 62 nm/RIU in a very small sensing area. This sensitivity would be reached even if only 1 μm² of the structure, around the cavity, is infiltrated with the solvent, which corresponds to an amount of solvent as small as 0.15 nl.

CONCLUSIONS

Resonant cavity structures which allow for multi-channel bio-detection are presented. Preliminary experimental demonstration of a double-channel device as a solvent sensor has been performed. After functionalizing the pore walls, these sensors will be capable of capturing and detecting bio-molecules, such as proteins. They are ideally suited to the simultaneous detection of multiple targets.

ACKNOWLEDGMENTS

This work is supported by the National Science Foundation (CBET 0730469). Device fabrication is performed in part at the Cornell NanoScale Facility, a member of the National Nanotechnology Infrastructure Network, which is supported by the National Science Foundation (Grant ECS 03-35765).

REFERENCES

1. Fan, X., White, I.M., Shopova, S.I., Zhu, H., Suter, J.D. and Sun, Y., 'Sensitive optical biosensors for unlabeled targets: A review', Analytica Chimica Acta 620, 8-26, (2008).
2. Homola, J., Yee, S.S. and Gauglitz, G., 'Surface plasmon resonance sensors review', Sensors and Actuators B 54, 315, (1999).
3. Guillermain, E., Lysenko, V., Orobtchouk, R., Benyattou, T., Roux, S., Pillonnet, A. and Perriat, P., 'Bragg surface wave device based on porous silicon and its application for sensing', Applied Physics Letters 90, 241116, (2007).
4. Luff, B.J., Wilkinson, J.S., Piehler, J., Hollenbach, U., Ingenhoff, J., and Fabricius, N., 'Integrated optical Mach-Zehnder biosensor', J Lightwave Technol 16(4), 583–592, (1998).
5. Rowe-Taitt, C.A., Hazzard, J.W., Hoffman, K.E., Cras, J.J., Golden, J.P. and Ligler, F.S., 'Simultaneous detection of six biohazardous agents using a planar waveguide array biosensor', Biosensors & Bioelectronics 15, 579-589, (2000).
6. Boyd, R. and Heebner, J.E., 'Sensitive disk resonator photonic biosensor', Applied Optics 40, 5742-5747, (2001).
7. Joannopoulos, J., Villeneuve, P.R. and Fan, S., 'Photonic crystals: putting a new twist on light', Nature 386, 143-149, (1997).
8. Skivesen, N., Têtu, A., Kristensen, M., Kjems, J., Frandsen, L.H. and Borel, P.I., 'Photonic-crystal waveguide biosensor', Optics Express 15, 3169-3176, (2007).
9. Lee, M. and Fauchet, P.M., 'Two-dimensional silicon photonic crystal based biosensing platform for protein detection', Optics Express 15, 4530-4535, (2007).
10. Lee, M.R. and Fauchet, P.M., 'Nanoscale microcavity sensor for single particle detection', Optics Letters 32, 3284-3286, (2007).
11. Mandal, S. and Erickson, D., 'Nanoscale optofluidic sensor arrays', Optics Express 16, 1623-1631, (2008).
12. Guillermain, E. and Fauchet, P.M., 'Multi-channel biodetection via resonant microcavities coupled to a photonic crystal waveguide', Proc. of SPIE 7167, (2009).
13. Bogaerts, W., Taillaert, D., Luyssaert, B., Dumon, P., Van Campenhout, J., Bienstman, P., Van Thourhout, D. and Baets, R., 'Basic structures for photonic integrated circuits in Silicon-on-insulator', Optics Express 12, 1583-1591., (2004).

Sensing and Detection on Chip – DNA

Mater. Res. Soc. Symp. Proc. Vol. 1191 © 2009 Materials Research Society 1191-OO12-04

Phosphate-Dependent DNA Immobilization on Hafnium Oxide for Bio-Sensing Applications

Nicholas M. Fahrenkopf[1], Serge Oktyabrsky[1], Eric Eisenbraun[1], Magnus Bergkvist[1], Hua Shi[2], Nathaniel C. Cady[1]
[1]College of Nanoscale Science & Engineering, University at Albany, Albany, NY 12203, USA
[2]Department of Biological Sciences, College of Arts and Sciences, University at Albany, Albany, NY 12222, USA

ABSTRACT

Hafnium(IV) oxide (HfO_2) has replaced silicon oxide as a gate dielectric material in leading edge CMOS technology, providing significant improvement in gate performance for field effect transistors (FETs). We are currently exploring this high-k dielectric for its use in nucleic acid-based FET biosensors. Due to its intrinsic negative charge, label-free detection of DNA can be achieved in the gate region of high-sensitivity FET devices. Previous work has shown that phosphates and phosphonates coordinate specifically onto metal oxide substrates including aluminum and titanium oxides. This property can therefore be exploited for direct immobilization of biomolecules such as nucleic acids. Our work demonstrates that 5' phosphate-terminated single stranded DNA (ssDNA) can be directly immobilized onto HfO_2 surfaces, without the need for additional chemical modification or crosslinking. Non-phosphorylated ssDNA does not form stable surface interactions with HfO_2, indicating that immobilization is dependent upon the 5' terminal phosphate. Further work has shown that surface immobilized ssDNA can be hybridized to complementary target DNA and that sequence-based hybridization specificity is preserved. These results suggest that the direct DNA-HfO_2 immobilization strategy can enable nucleic acid-based biosensing assays on HfO_2 terminated surfaces. This work will further enable high sensitivity electrical detection of biological targets utilizing transistor-based technologies.

INTRODUCTION

Much progress has been made in using field effect transistors (FETs) as transduction elements for biomolecular sensors. Different strategies and architectures have been used, such as extended gate FETs [1], ion-sensitive FETs [2], enzyme FETs [3], high electron mobility transistors (HEMTs) [4], organic thin film transistors [5], dielectric modulating FETs [6], "nanobelt" FETs [7], and nanotube based FETs [8], among others. Deoxyribonucleic acid (DNA) is an attractive biomolecular target for FET-based sensors due to the high specificity of DNA hybridization events, and the intrinsic negative charge associated with the sugar-phosphate backbone. Indeed, research has shown that changes in the concentration of DNA immobilized on the gate region of an FET [9] are electrically detectable, and that the immobilization and hybridization events can both be sensed through current/voltage (I/V) measurements [10]. DNA-based FET biosensors mainly consist of single stranded DNA (ssDNA) "probe" immobilized on the gate region of the FET. Hybridization of this probe with target (complementary) ssDNA increases the amount of charge localized on the gate region. Typical methods to immobilize oligonucleotides onto metal oxides include chemical linkage to surface layers deposited via silanization [2, 10, 11], or to self-assembled monolayers deposited via gold-thiol interactions [4,

6]. These methods introduce additional process steps and increase the distance between the probe/target DNA and the gate of the FET. Recent research on the detection of DNA hybridization via optical methods has exploited the phosphate group that terminates the 5' end of DNA and the coordination of this phosphate group with Group IV transition metals, namely zirconium [12, 13]. At least one group has also shown that phosphate terminated organic molecules will coordinate to zirconium oxide and hafnium oxide [14]. In this work, we sought to directly immobilize probe ssDNA to hafnium oxide as a platform for future efforts in FET-based biosensors utilizing hafnium oxide as the gate dielectric.

EXPERIMENTAL METHODS

Hafnium oxide (HfO_2) films, approximately 15nm thick, were prepared by atomic layer deposition (ALD) onto silicon wafers using an Aixtron/Genus Stratagem^{300} 300-mm wafer ALD cluster tool. Tetrakis ethylmethylamido hafnium (TEMAH) and ozone were employed as reactants. Deionized water (dH_2O) was prepared by ion exchange to a resistivity of 18.2 MΩ-cm. All chemicals were commercially available and used as received unless noted otherwise. Tris-EDTA (TE) buffer, pH 8.0 was obtained from National Diagnostics (Atlanta, GA). Sulfuric acid, hydrogen peroxide (30%), and 20x SSC buffer (0.3M sodium citrate, 3M NaCl, pH 7.0) were obtained from Sigma-Aldrich (St. Louis, MO). PicoGreen fluorescent dye was purchased from Invitrogen (Carlsbad, CA) and diluted into TE buffer 1:400 (v/v) prior to use. Bovine serum albumin (BSA) was obtained from New England Biolabs (Ipswich, MA) and diluted to 0.1μg/μL in TE prior to use. Isopropyl alcohol was purchased from Fisher Scientific (Pittsburgh, PA). Single-stranded DNA oligonucleotides were purchased from Integrated DNA Technologies (Coralville, IA) having the following sequences: probe - 5' TCACCATTAGTACC AGAATCAGTAATTC; complement - 5' **G**AATTACTGATTCT**G**GTACTAATGGTGA; non-complement - 5' GCGTCGTACTATTGAAAAGCTGTCT; complement with one mismatch, 5' AATTACTGATTCT**A**GTACTAATGGTGA. Probe DNA was supplied from IDT with a 5' terminal, a 3' terminal or no terminal phosphate ("probe-5P", "probe-3P" and "probe", respectively). Upon receipt, all oligonucleotides were reconstituted to 500 mM with filter sterilized dH_2O and stored at -20°C.

DNA Patterning

HfO_2 surfaces were cleaned prior to printing with four different protocols: 1) immersion in piranha solution (30% H_2O_2:98%H_2SO_4, 1:2 (v/v)) for 5 min and rinsed three times with 50 mL dH2O before being blown dry with nitrogen gas; 2) cleaned for 3 min at 15W in a Harrick PDC-32G Plasma Cleaner (Ithaca, NY); 3) cleaned for 20 minutes in a BioForce UV/Ozone ProCleaner (Ames, IA); or 4) cleaned by immersion in isopropyl alcohol (IPA) for 15 min in an ultrasonic bath and blown dry with nitrogen gas. Some samples were not cleaned prior to DNA patterning and were used "as received" after HfO_2 deposition.

ssDNA probe DNA was patterned onto HfO_2 coated surfaces using a BioForce Nano eNabler (NeN) (BioForce Nanosciences, Ames, IA) [15]. Briefly, the ssDNA to be printed (probe, probe-5P, or probe-3P) was diluted to 5μM in TE buffer with 10% (v/v) glycerol to prevent evaporation. The DNA solution was loaded into the reservoirs of BioForce 30μm surface patterning tools (SPTs) that had been treated in a BioForce ProCleaner UV/ozone cleaner for 20 min. SPTs loaded with sample fluid were mounted in the NeN for printing directly onto the various HfO_2 substrates. The NeN chamber was held at 70% relative humidity during printing which resulted in deposited spots with an average diameter of 30μm. SPTs were cleaned and reloaded to enable printing of additional DNA sequences onto an individual sample. Following printing, the samples were allowed to incubate for 12 hr at room temperature.

DNA Hybridization

Patterned HfO_2 surfaces were rinsed three times with 0.8 mL TE buffer before being blocked in a solution of 0.1 mg/mL bovine serum albumin (BSA) in 0.8 mL TE buffer for 20 min. Previous studies have shown that pre-blocking surfaces with BSA improves the signal to noise ratio [15] by reducing non specific interactions between target DNA and the surface. HfO_2 surfaces were then rinsed three times with 0.8mL TE buffer and immersed in a hybridization solution 0.8 mL of 5x SSC buffer with 0.1 μM target DNA. The hybridization solution was incubated at 70°C for 5 min and allowed to cool to room temperature before being rinsed three times with 0.8mL TE buffer. The samples were then immersed in 0.8mL of 1:400 diluted PicoGreen dye in TE for 20 min for staining, followed by rinsing with TE as above. The surfaces were immediately covered with glass coverslips and imaged using a Nikon Eclipse 80i epifluorescence microscope (Nikon USA, Melville, NY) with 490/520nm excitation/emission.

Fluoresce Intensity Measurements

Fluoresce intensity analysis of the HfO_2 surfaces was performed using ImageJ, the freely accessible software from the National Institutes of Health (NIH) [16]. For each sample, the average intensity of three different spots was taken, along with equivalent background areas next to each spot. The fluorescence signal for each spot was determined by normalization to the background intensity: Signal = [(Spot intensity / Background intensity) – 1]. The normalized signal for each spot was then averaged and the standard deviation of these values was calculated using Microsoft Excel.

RESULTS AND DISCUSSION

DNA immobilization through terminal phosphate groups was evaluated by printing probe, probe-5P and probe-3P DNA onto a sample surface. The samples were hybridized with complementary DNA, stained with PicoGreen and imaged as described above. Through this approach, we evaluated where ssDNA had been immobilized to the surface and successfully hybridized with target DNA. As shown in Figure 1, the oligonucleotides were printed in a checkerboard-like pattern to clearly differentiate between the different probes (phosphorylated or non-phosphorylated). Figure 2 compares hybridized and PicoGreen stained spots of probe, probe-5P and probe-3P DNA. In addition, probe-5P DNA was printed and hybridized with non-complementary DNA and complementary DNA with one mismatch. Three representative spots from each sample (and background areas) were measured for average signal intensity using ImageJ, as described above. The results of these fluorescence measurements are shown in Figure 2.

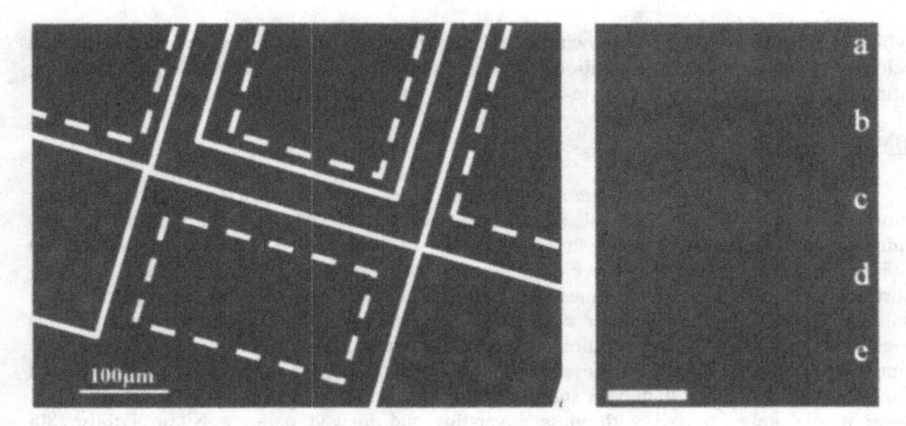

Figure 1. Left: Optical fluorescence image of ssDNA patterned onto oxygen plasma cleaned HfO_2, hybridized with complementary DNA, and stained with PicoGreen. Areas bounded by solid lines are where probe-5P was printed. Areas bounded by dashed lines indicate where probe DNA (no phosphate) was printed. **Right:** Representative spots of probe-5P DNA hybridized with complementary target after being printed on HfO_2 that was **(a)** used as received, **(b)** piranha cleaned, **(c)** oxygen plasma cleaned, **(d)** UV/ozone cleaned, or **(e)** IPA/sonicator cleaned (scale bar 50um).

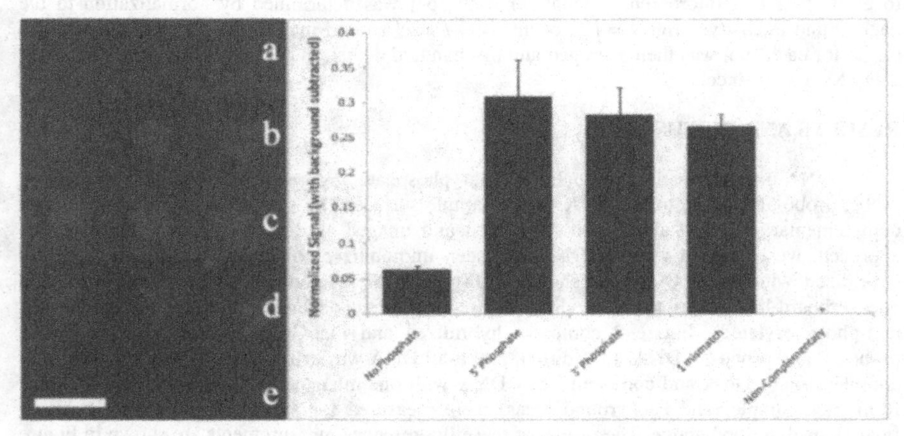

Figure 2. Left: Representative spots from ssDNA printed on piranha cleaned HfO_2, scale bar 50um. a) Probe DNA with no terminal phosphate, (b) probe with 5' prime phosphate, and (c) probe with 3' phosphate. Each of these (a-e) were hybridized with complementary target DNA. Probe DNA with 5' phosphate hybridized with single mismatch (d) or non-complementary DNA (e). **Right:** Normalized fluorescence signal for each experimental condition (error bars = 1 standard deviation).

These data illustrate that phosphate terminated ssDNA can be patterned onto HfO_2 using the NeN system and that the ssDNA is firmly anchored to the surface, which allow hybridization of complementary DNA. Immobilization is dependent upon terminal phosphate modification of the DNA since probe without a terminal phosphate group was shown to have little attachment to the surface (Figure 1, left). Hybridization with target DNA was shown to be sequence specific, where non-complementary target failed to hybridize, and a single-mismatch target had reduced hybridization indicated by the lower fluorescent intensity after PicoGreen staining (Figure 2).

To evaluate the effectiveness of various surface preparations, probe DNA was printed onto samples of HfO_2 that had been cleaned with different methods: piranha, oxygen plasma, UV/ozone, IPA/sonication, and uncleaned "as received". Since most CMOS devices contain metals and oxides that would not withstand strong acid based treatments (i.e. the piranha clean), alternative methods were sought. The right side of Figure 1 shows the differences in spot morphology and intensity for each of the different treatments. Although DNA was successfully immobilized and hybridized on surfaces prepared by each method, piranha-based cleaning resulted in the optimal morphology and spot intensity. Treatments (c) and (d) resulted in surfaces with inconsistent spot printing, and while (e) produced consistent spot patterns, the intensity was less than that of other treatments. The exact mechanism of phosphate mediated immobilization of DNA to HfO_2, and the effects of various surface treatments on the immobilization are areas of ongoing research. The variation of spot morphology and fluorescence intensity are likely the result of a combination of factors, such as physical/chemical modification of the surface interface during the cleaning process and the efficiency of surface contaminant removal.

CONCLUSIONS

Single-stranded probe DNA can be directly immobilized onto HfO_2 surfaces without the need for additional chemical cross-linkers or complex surface preparation. Our work shows that immobilized DNA is available for sequence specific hybridization to target DNA. Further we show that a terminal phosphate group, either on the 3' or 5' DNA terminus, is necessary for immobilization. The treatment of the surface before printing influences the printed spot morphology and the degree of immobilization. Further research is required to better understand the specific mechanisms of immobilization of nucleic acids on HfO_2, and additional studies need to be conducted to characterize the selectivity of hybridization to mismatched target, and sensitivity to target concentration.

ACKNOWLEDGMENTS

The authors wish to acknowledge and thank BioForce Nanosciences, the College of Nanoscale Science and Engineering/College of Arts and Sciences Nanobio Challenge Grant, Dr. Natalya Tokranova, Zachary Rice, and Sarah Goulet.

REFERENCES

[1] M. Kamahori, Y. Ishige, and M. Shimoda, "Enzyme Immunoassay Using a Reusable Extended-gate Field-Effect-Transistor Sensor with a Ferrocenylalkanethiol-modified Gold Electrode," *Analytical Sciences*, vol. 24, pp. 1073-1079, 2008.

[2] D. Goncalves, D. M. F. Prazeres, V. Chu, and J. P. Conde, "Detection of DNA and proteins using amorphous silicon ion-sensitive thin-film field effect transistors," *Biosensors and Bioelectronics,* vol. 24, pp. 545-551, 2008.
[3] S. Migita, K. Ozasa, T. Tanaka, and T. Haruyama, "Enzyme-based Field-Effect Transistor for Adenosine Triphosphate (ATP) Sensing," *Analytical Sciences,* vol. 23, p. 4, 2007.
[4] S. J. P. B. S. Kang, J. J. Chen, F. Ren, J. W. Johnson, R. J. Therrien, P. Rajagopal, J. C. Roberts, E. L. Piner, and K. J. Linthicum, "Electrical detection of deoxyribonucleic acid hybridization with AlGaN/GaN high electron mobility transistors," *Applied Physics Letters,* vol. 89, pp. 122102-122104, 2006.
[5] Q. Zhang and V. Subramanian, "DNA hybridization detection with organic thin film transistors: Toward fast and disposable DNA microarray chips," *Biosensors and Bioelectronics,* vol. 22, pp. 3182-3187, 2007.
[6] X.-J. H. HYUNGSOON IM, BONSANG GU AND YANG-KYU CHOI,"A dielectric-modulated field-effect transistor for biosensing," *Nature Nanotechnology,* vol. 2, p. 5, 2007.
[7] Y. Cheng, P. Xiong, C. S. Yun, G. F. Strouse, J. P. Zheng, R. S. Yang, and Z. L. Wang, "Mechanism and Optimization of pH Sensing Using SnO2 Nanobelt Field Effect Transistors," *Nano. Lett.,* vol. 8, pp. 4179-4184, 2008.
[8] H. Yoon, J. Kim, N. Lee, B. Kim, and J. Jang, "A Novel Sensor Platform Based on Aptamer-Conjugated Polypyrrole Nanotubes for Label-Free Electrochemical Protein Detection," *Chembiochem,* vol. 9, pp. 634-641, 2008.
[9] G. Xuan, J. Kolodzey, V. Kapoor, and G. Gonye, "Characteristics of field-effect devices with gate oxide modification by DNA," *Applied Physics Letters,* vol. 87, pp. 103903-103905, 2005.
[10] T. Sakata, M. Kamahori, and Y. Miyahara, "DNA Analysis Chip Based on Field-Effect Transistors," *Japanese Journal of Applied Physics,* vol. 44, pp. 2854-2859, 2005.
[11] G. S. B. Baur, J. Hernando, O. Purrucker, M. Tanaka, B. Nickel, M. Stutzmann and M. Eickhoff, "Chemical functionalization of GaN and AlN surfaces," *Applied Physics Letters,* vol. 87, pp. 263901-263903, 2005.
[12] B. Bujoli, S. M. Lane, G. Nonglaton, M. Pipelier, J. Leger, D. R. Talham, and C. Tellier, "Metal Phosphonates Applied to Biotechnologies: A Novel Approach to Oligonucleotide Microarrays," *Chem. Eur. J.,* vol. 11, pp. 1980-1989, 2005.
[13] K. S. Rao, S. U. Rani, D. K. Charyulu, K. N. Kumar, H. Lee, and T. Kawai, "A novel route for immobilization of oligonucleotides onto modified silica nanoparticles," *Analytica Chimica Acta,* vol. 576, p. 7, 2006.
[14] M. L. Jespersen, C. E. Inman, G. J. Kearns, E. W. Foster, and J. E. Hutchison, "Alkanephosphonates on Hafnium-Modified Gold: A New Class of Self-Assembled Organic Monolayers," *J. AM. CHEM. SOC.,* vol. 129, p. 5, 2007.
[15] X. Xu, V. Jindal, F. Shahedipour-Sandvik, M. Bergkvist, and N. C. Cady, "Direct immobilization and hybridization of DNA on group III nitride semiconductors " *Applied Surface Science,* vol. 255, pp. 5905-5909, 2009.
[16] W. S. Rashband, "ImageJ," Bethesda, MD: US National Institutes of Health, 1997-2008.

Printed in the United States
By Bookmasters